LB-III-46

European Commission

Directorate-General for Research

D1657336

Wind erosion on agricultural land in Europe

Research results for land managers

Edited by
Andrew Warren
University College London

Geographisches Institut
der Universität Kiel
ausgesonderte Dublette

Geographisches Institut
der Universität Kiel

Environment and Sustainable Development Programme

2002

EUR 20370

Geographisches Institut
der Universität Kiel
ausgesonderte Dublette

EUROPEAN COMMISSION

Directorate-General for Research

LEGAL NOTICE: Neither the European Commission nor any person acting on behalf of the Commission is responsible for the use which might be made of the following information.

A great deal of additional information on the European Union is available on the Internet.
It can be accessed through the Europa server (http://europa.eu.int).

Cataloguing data can be found at the end of this publication.

Luxembourg: Office for Official Publications of the European Communities, 2003

ISBN 92-894-3958-0

© European Communities, 2003
Reproduction is authorised provided the source is acknowledged.

Printed in Belgium

PRINTED ON WHITE CHLORINE-FREE PAPER

Wind Erosion on Agricultural Land in Europe

Contents:

FOREWORD

Wind erosion is also a European phenomenon. It has been addressed within the EC Environment Research Programme in the context of the topic "Land Degradation and Desertification".

This publication is one of the outcomes of three EC funded projects. Two of these are research projects: WELSONS (Wind Erosion and Loss of Soil Nutrient in semi-arid Spain) and WEELS (Wind Erosion on European Light Soils), and one was an accompanying measure ECOWEAL (European Conference on Wind Erosion on Agricultural Land)

There is always a need, beside the basic scientific results presented in the literature, to disseminate, to a wider audience, relevant research results and information. This publication covers therefore the practical or applied implications of these projects along with a wider survey of the relevant literature including the Australian and US experience. The objective is to present the main issues which need to be considered by policy-makers and those who direct research on these matters in Europe.
It concludes with the summary of a discussion held at the end of the ECOWEAL conference, in which participants from across Europe brought together ideas about the steps forward for European research into and policy about wind erosion.

In the context of the recent Commission's Communication "Towards a thematic strategy for soil protection" this booklet is timely. I am convinced that it will contribute to highlight the problems, provide a comprehensive review and stimulate further interest in this field of research

Finally, I wish to thank the co-ordinators namely Pierre Gomez, Geert Sterk and Andrew Warren and their teams involved in these projects as well as my colleague Denis Peter for his efforts to support and efficiently manage this area of research.

Anver Ghazi
Head of Unit
Biodiversity and Global Change

1. Introduction

Andrew Warren, University College London and Lars Bärring, Lund Universit

This booklet is intended to describe, where, when and under what circumstances of lanc wind erosion can occur, what effects it has, on-site and off-site, and what can be done about ıt. It has chapters on what has been achieved in the United States and Australia in these respects. The booklet is aimed at rural land managers and policy makers. This introduction puts the problem in historical perspective and gives a brief introduction to the processes involved.

Wind erosion as a problem

Wind erosion creates many problems on Europe. In northern Europe the problem is severe only on light, sandy soils, but given the extent of the sandy Pleistocene glacial outwash from which these soils are mostly developed, wind erosion itself is not a trivial issue. Wind erosion also occurs on more silt- or clay-rich soils in the drier parts of southern Europe, but the problem is less well researched, and probably less extensive or intense. The problems that wind erosion brings, in both these settings are: loss of crops, pollution (as from dust and pesticides) and jeopardised sustainability.

Wind erosion in European history

The problems have been known for millennia. Archaeological evidence shows that wind erosion, exacerbated by cultivation and grazing, has occurred since the Neolithic. Over the years since then, whole villages have been eradicated by wind erosion in Scotland, The Netherlands and Denmark (Edlin 1976: Heidinga 1984; Skarregaard 1989). These early incidents may have taken place under somewhat different climatic conditions than exist today (Vandenburghe 1993), but there is little doubt that they were all induced by changes in land use.

The archaeological evidence shows a similar pattern in eastern England from the Neolithic, and here, as elsewhere, wind erosion is known to have persisted up to the 17th century, when an early paper in the *Proceedings of the Royal Society* reported the problem (Wright 1669), and the 19th, when it was graphically described by the writer Cobbett in his *Rural Rides* (1830). Cobbett asked a farmer whether his farm was in Suffolk or Essex: "it depends on which way the wind is blowing", was his reply. Serious wind erosion was also recognised in the 17th-century *Sand Boards* of the Veluwe in the Netherlands and the protracted efforts by the Danish *Hedesaellskabet.* In the middle of the eighteenth century von Linnaeus, the celebrated Swedish botanist, described wind erosion in Scania at a time when the agricultural system was in a crisis brought on by deforestation and the cultivation of new land by a growing population (Linnaeus 1751). Traces of the crisis can be seen today as fossil dunes and in the pine forests planted to protect vulnerable land in Scania. Notwithstanding these efforts, the *Swedish Committee on Erosion* estimated in 1950 that 35 000 ha were still subject to erosion, mostly of them in Scania. In France, Brémontier's classic efforts, at the command of Napoleon, to reclaim the Quaternary sands of the Landes in Aquitaine, starting in 1787, are another example of major attempts to combat a very similar problem (Brémontier 1797).

Mechanisation, increases in field size and contract farming are probably exacerbating the rate of soil loss. Despite extensive research in control methods, there are few good data on either damage or the economic efficiency of the control measures, let alone criteria for applying laws and codes. Yet what data there are do suggest a major problem. For example, the direct cost only for the resowing of sugar beet after one single storm in May 1984 was estimated to be approximately 1.5 million ECU for sugar beet fields in Scania alone.

The factors that influence wind erosion

The scientific study of wind erosion has a long history in Europe and the United States. In Europe the pioneers built on the work of the aerodynamicists Prandtl (e.g.1935) and von Kármán (e.g.1934), and the engineer Shields (1936). This work was first applied to the work of the wind by the engineer Bagnold (1941). In the United States and Canada one of the greatest pioneers was Chepil (1957). There are many summaries of this work and its more recent development, as in Greeley and Iversen (1989), Cooke, Warren and Goudie (1992); and Livingstone and Warren (1996). What follows is a simplified introduction to the process of wind erosion, intended as an background to the discussion later in this booklet.

Wind erosion occurs when three conditions are met: the wind is strong enough, the soil surface is susceptible enough and there is no surface protection by a crop, mulch or snow. These conditions are usually placed in two categories:

Erodibility expresses the inherent potential of a soil to erode: the soil's resistance to the wind. Properties that influence erodibility are:

- the particle size of the soil (in general, coarse, sandy soils are most susceptible because they dry quickly and are poorly aggregated), and
- its organic matter content (which also influences aggregation).
- the soil moisture content of the surface, or more precisely the soil moisture regime which influences the surface drying of a soil and thus the duration of susceptibility.

Erodibility is highly variable within and between soils, and from one time to another.

Erosivity describes the eroding capacity of the wind. It expresses the transfer of momentum from the wind to the surface, or the force that the wind exerts on the soil.

The factors that control erosivity and erodibility, respectively, are a mixture of natural factors and factors influenced by human activities (Figure 1).

Wind conditions near the surface depend on the weather and surface conditions. When the surface is not too rough, as in open and gently undulating agricultural land, the increase in the velocity of the wind above the surface is a function of the undisturbed wind and the roughness of the surface. At about 50 m above such a surface the wind speed is mainly determined by weather (temperature and atmospheric pressure differences). But closer to it, the wind is slowed by roughness, which is a function of three factors:

Figure 1: Conceptual outline of principal relationships between factors important for wind erosion. The three overall factors landscape, weather and natural soil conditions are placed in separate boxes. The fourth overall factor, farmer's actions are highlighted as yellow text boxes. The centrepiece of this diagram is the relation between the erosivity (friction velocity of the wind) and the actual erodibility (threshold friction velocity) of the soil surface. The friction velocity of the wind is a function of the large-scale wind speed and the effective roughness that integrates roughness from micro- to macro-scale. Erodibility is a function of soil particle size distribution, soil aggregate structure and stability, surface crusting, and soil surface moisture. These factors are in turn determined by physical, chemical and biological processes in the soil.

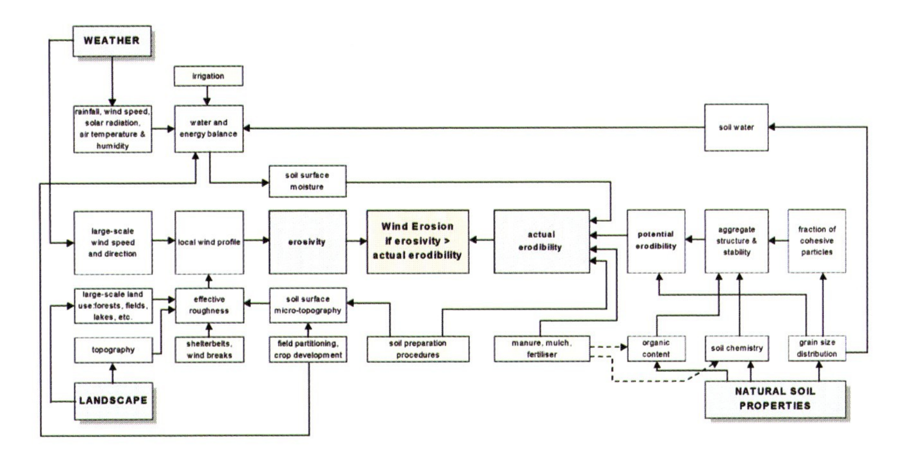

- the roughness of the bare, uncovered soil, itself controlled only by the soil properties (particle size and aggregation); large aggregates protect erodible grains from wind shear and trap moving particles;
- the nature of cultivation (type of tillage and the direction of tillage in relation to wind direction, etc.), which are another form of roughness; roughness induced by tillage is a good method of controlling wind erosion; and
- the roughness produced by the vegetation cover related to the height and density of the crop; annually sown crops do not provide protection at some times of the year; besides the sowing date, the length of time taken for each crop to grow to a sufficient size is a fundamental factor. Crop residues left on the soil surface are also important in protecting it.

Another important process is the 'saturation' of the wind with soil particles as it blows over a susceptible surface. The length of the field required for the wind to attain its maximum capacity depends on its velocity and the composition of the soil. Beyond the saturation point, the quantity of eroded soil in the wind remains virtually constant. The distance needed for the wind to attain 63.2 % of its transport capacity is called the *critical field length (in the wind direction)*. It is said to be less than 150 m for most surfaces and soils. Knowing the critical field length is an important step in effective wind erosion control. By decreasing or breaking up the length of a field, erosion can be reduced.

The interaction of the various factors determines the potential and actual erodibility of a particular site. The potential erodibility in this context depends on soil and surface character, and these can be influenced by cultivation measures.

References

Bagnold, R.A. 1941. *The physics of blown sand and desert dunes*, Methuen, London, 265 pp. [2nd printing, 1954, 3rd 1960

Brémontier, N.T. 1797. *Memoire sur les dunes, et particulierement sur celles qui se trouvent entre Bayonne et la pointe de Grève, a l'embouchure de la Gironde*, L'imprimérie de la République, Paris, 73 pp.

Chepil, W.S. 1957. Erosion of soil by wind, *Soil, 1957 Yearbook of Agriculture*, United States Department of Agriculture, 308-314.

Cobbett, W. 1967. *Rural rides*, Penguin, Harmondsworth, 533 pp. [1st edition 1830

Cooke, R.U., Warren, A. and Goudie, A.S. 1992. *Desert Geomorphology*, University College Press, London, 512 pp.

Edlin, H.L. 1976. The Culbin Sands, in *Reclamation*, Eds. Lenihan, J. and Fletcher, W.W., Blackie, Glasgow, 1-31.

Greeley, R. and Iversen, J.D. 1985. *Wind as a geological process on Earth, Mars, Venus and Titan*, Cambridge University Press, Cambridge, 333 pp.

Heidinga, H.A. 1984. Indications of severe drought during the 10th century A.D. from an inland dune area in the central Netherlands, *Geologie en Mijnbouw*, **63** (3), 241-248.

Linnaeus, C., 1751. *Carl Linnaei skånska resa, på höga öfwerhetens befallning förrättad år 1749: med rön och anmärkningar uti oeconomien, naturalier, antiquiteter, seder, lefnads-sätt*, Stockholm.

Livingstone, I. and Warren, A. 1996. *Aeolian geomorphology: an introduction*, Addison-Wesley Longman, Harlow, 211 pp.

Prandtl, L. 1935. The mechanics of viscous fluids, in *Aerodynamic theory, **III***, Ed. Durand, W.F., Springer-Verlag, Berlin, 34-208.

Shields, A. 1936. *Anwendung der Ähnlichkeitsmechanik und der Turbulenzforschung auf die Geschiebebewegung, Mitteilungen der preussischen Versuchsandstalt für Wasserbau und Schiffbau*, **26**, Berlin, 26 pp.

Skarregaard, P. 1989. Stabilisation of coastal dunes in Denmark, in *Perspectives in coastal dune management*, Eds. van der Meulen, F., Jungerius, P.D. and Visser, J., SPB Scientific, The Hague, 151-162.

Vandenberghe, J. 1993. Changing conditions of aeolian sand deposition during the last deglaciation period, in *Late Vistulan (= Weichselian) and Holocene aeolian phenomena in North and West Europe*, Ed. Kozarski, S., *Zeitschrift für Geomorphologie, Supplement Band*, **90**, 193-207.

von Kármán, T. 1934. Turbulence and skin friction, *Journal of Aeronautical Science*, **1**, 1-20.

Wright, T. 1669. A curious and exact relation of a sand cloud which hath lately overwhelmed a great tract of land in the county of Suffolk, *Philosophical Transactions of the Royal Society of London*, **3**, 722-725.

2. Wind Erosion in Europe: Where and When

Jens Groß, now at University of Göttingen and Lars Bärring, Lund University

Introduction

Wind erosion is not a major problem in most of Europe, but it can cause severe damage in some areas. Some of the problems, such as in the creation of dust, occur mainly in the east and south (EEA, 1998), where there is a combination of susceptible soils, dry and hot conditions, and particular cultivation practices. Wind erosion is also a problem in the temperate climates of northwestern Europe where it primarily affects the light soils derived from Quaternary glacial outwash. In these northwestern regions, losses of soil in single-events in the spring may exceed 5 t ha^{-1} once in 10 years and may reach 40 t ha^{-1} in one event (Beinhauer and Kruse 1994). Most of the fields that are affected are those that are used for intensive agriculture, and here the ecological and economic consequences can be serious and may be irreversible. The risk of wind erosion has increased decisively within the last few decades, following the intensification of agriculture, the enlargement of fields (EEA, 1998), and the increasing cultivation of crops that are associated with wind erosion, such as maize, in Lower Saxony, or sugar beet in southern Sweden. In such areas, agricultural practices may also play a major role in controlling the factors that influence the location and rate of wind erosion.

The increase in the environmental and economic problems that result from wind erosion has led to a demand for prediction systems that provide policy makers, producers and other potential stakeholders with information about fields under threat, actual soil losses and the effects of modified agricultural management practices on the spatial and temporal occurrence of wind crosion.

This chapter describes the general pattern of occurrence of wind erosion in Europe and the major contributing factors. It then outlines methods to assess the distribution and possible extent of wind erosion at different spatial scales. Finally it looks at the temporal variation in the occurrence of wind erosion, from the hourly to the millennial scale.

Distribution of wind erosion in Europe as a whole

The European Environment Agency (EEA, 1998) provides a European-wide assessment of the distribution and severity of wind erosion (Figure 1) on the basis of published observations (Van Lynden, 1995) and measurements in the field. The extent of wind erosion is assessed in terms of loss of topsoil. The provisional nature of this assessment, based as it is on very poor data (which are much poorer for some countries than others), must be emphasised.

The most extensive and most severe wind erosion is mapped in southeastern Europe, in Romania, the Ukraine and Russia. Moderate wind erosion is also thought to occur in the Czech Republic and in parts of France, the UK, and Hungary. Wind Erosion is also indicated as an important problem in Iceland. The map shows a belt of light wind erosion in the Quaternary deposits of the northern European plain, which extends from Belgium to beyond

Poland. According to Oldeman *et al.* (1991, cited by the EEA, 1998) 42 million ha or 4 % of the EU area is affected by wind erosion.

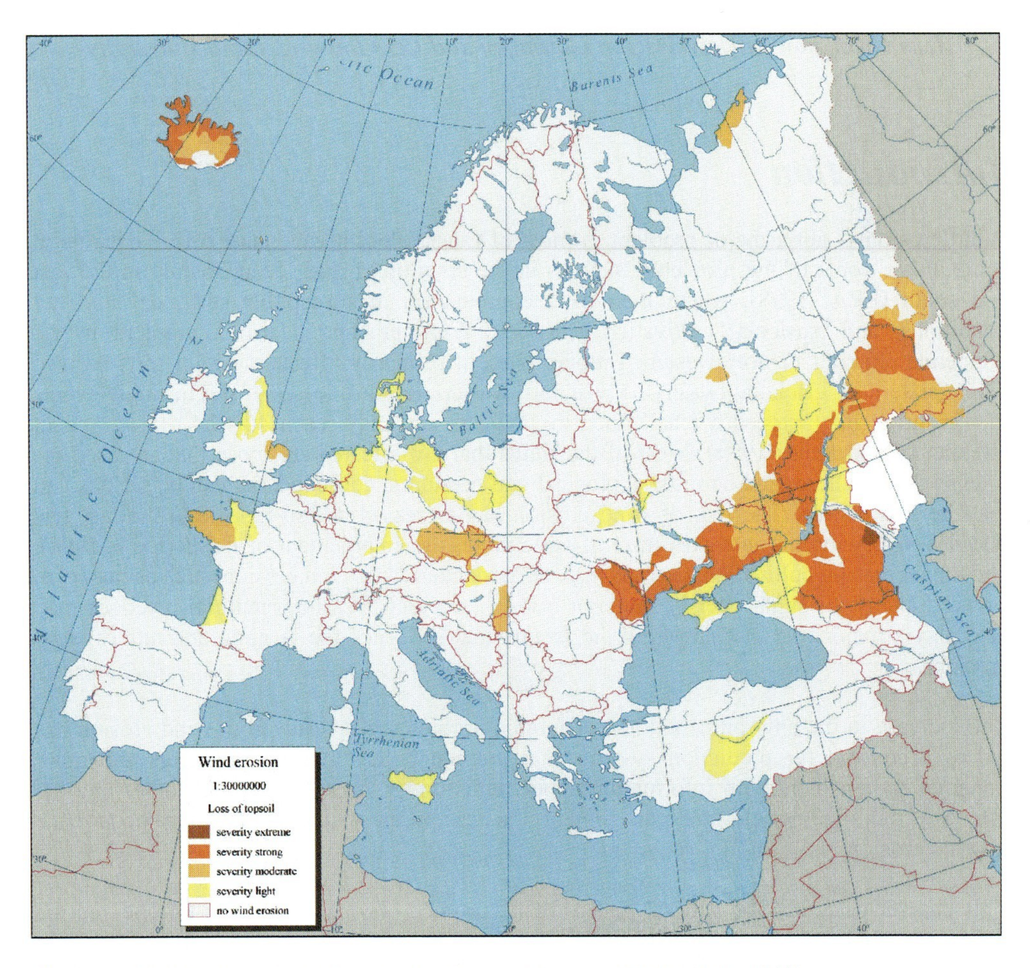

Figure 1: Distribution of wind erosion in Europe (Source: ISRIC, EEA 1998)

Recent investigations in the framework of EU-Projects (WEELS - '*Wind Erosion on European Light Soils*' and WELSONS -'*Wind Erosion and Loss of Soil Nutrients in Semi-Arid Spain*') suggest that the potential and actual affected area is probably much larger than shown on Figure 1, for within regions shown on Figure 1 to be only slightly affected, there are 'hot spots' where erosion is more serious. Large areas in East Anglia, Lower Saxony (Germany), and the southern part of Sweden are temporarily affected by significant wind erosion events. The WEELS project has estimated rates of 21 t ha^{-1} yr^{-1} over a 30-year period in a part of East Anglia, a rate that compares closely with the estimated rates of water erosion in other parts of England (see Figure 2).

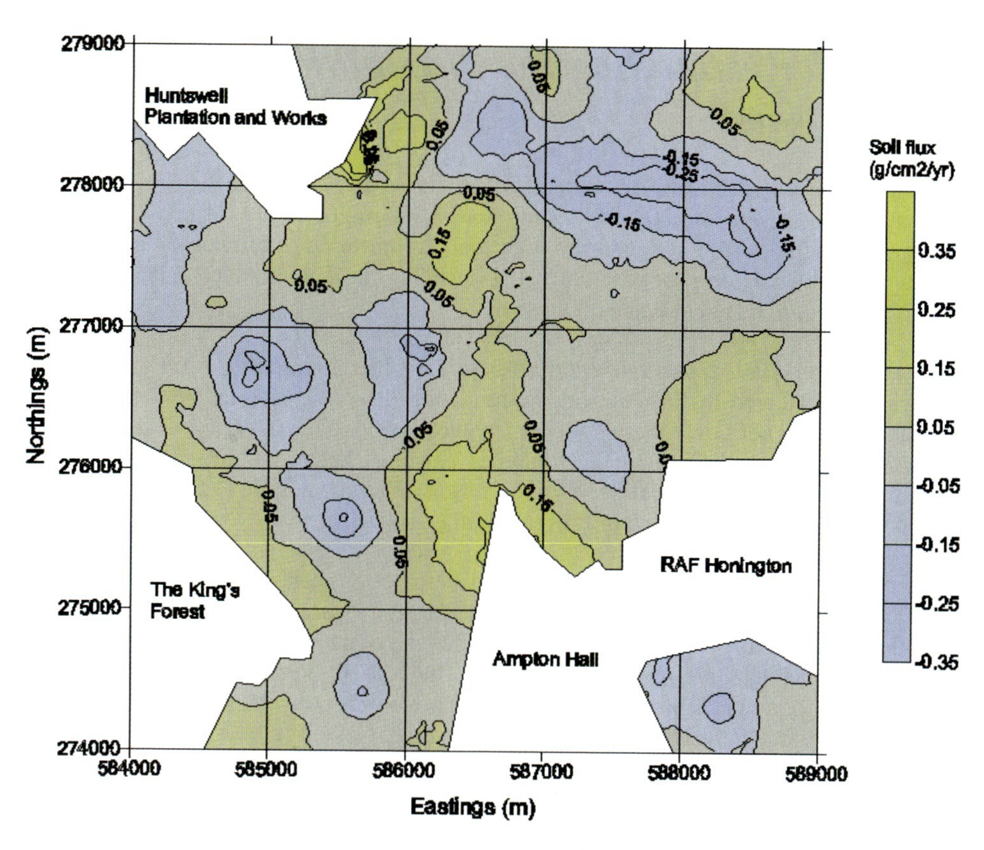

Figure 2: Net soil flux (+ erosion and - deposition g cm^{-2} yr^{-1}) over 30 years in the WEELS study area (25 km^2) in East Anglia. The data are derived from measurements of ^{137}Cs and mapped every 50 m by geostatistical estimation. The coordinates are those of the Ordnance Survey, in metres. No samples were taken from the blank areas that are covered by a military airfield, a farmer who prohibited access and forest.

Apart from their sandy soils, the open nature of these landscapes and their gentle slopes are important factors that encourage wind erosion. Their susceptibility to wind erosion has also increased as fields have been enlarged. The affected area in Europe is estimated to be about 2 million ha of land in Lower Saxony in Germany, 97 000 ha in the Netherlands, about one million ha in the western part of Denmark, 170 000 ha in Sweden and 260 000 ha in the UK (Riksen and De Graaff, 2001). In southern Sweden, many arable fields have already been removed from production because of a substantial decline of the soil productivity over recent years that has been caused by wind-erosion (Jönsson, 1992). The situation is also very critical in Eastern Europe, where land in former state-controlled farms has suffered considerable erosion on very large arable fields (EEA, 2000). Large areas in the northern and western parts of Poland have already been removed from production (Veen *et al.*, 1997).

Wind erosion is also a potential hazard in the Mediterranean part of Europe, in particular on unprotected land. The extent of wind-induced soil degradation has recently been investigated in Central Aragón, northeastern Spain, in the framework of the EU-Project WELSONS (Sterk *et al.* 1999).

Methods for estimating the distribution and severity of wind erosion.

If we want to plan for sustainability, we must determine where and under which conditions damage by wind erosion is likely to occur and what might be its effects. Prediction technology is required to provide planners with tools to locate potentially erodible land and producers with guidance on how to protect it. Depending on their primary objective, users will require answers to different kinds of question.

Identification of problem areas on regional scale. Data that might be relevant to assessing wind erosion are gathered by different organisations for different purposes. Few are directly relevant and most do not cover the whole geographical area of interest (EEA 2000). One of the most advanced approaches, which combines existing data from different sources to provide information for policy relating to wind erosion on regional scale (1:50.000) has been developed by the Geological Survey of Lower Saxony.

In Lower Saxony regional wind erosion risk is located by a method that utilizes existing statewide soil data (e.g. data on the German Soil Taxation system) and land-use statistics (Thiermann *et al.* 2000). On the basis of soil particle-size characteristics, the topsoil layer is classified into six potential erosion risk classes, from 'no risk' to 'very high risk'. Soils with a high proportion of clay and silt are classified as 'non-erodible' and sandy soils with a high proportion of fine sand are in the highest wind erosion risk class. The next step uses land use statistics (provided by the Department of Statistics), which classifies the typical crop types grown in northern Germany according to the plant coverage during the wind erosion season in spring (most events occur in March to May, see below). The best protection is provided by pasture, which has year-round ground cover, whereas because of to their late sowing date, maize or vegetables do not provide sufficient protection before summer. The combination of these factors reveals, for each community, the proportions of highly erodible fields that are sown to erosion-sustaining crops. The determination of these 'hot spots' on a broader scale makes it possible to implement wind erosion measures more efficiently and to concentrate further investigations on relevant regions.

Determination of potential wind erosion risks on field scale. Once the main wind erosion risk areas have been located, the implementation of control measures requires the determination of the potential wind erosion risks for particular fields.

The Geological Survey of Lower Saxony has developed a suitable tool for this purpose (Thiermann *et al.* 2000). Potential wind erosion risks are determined by means of a method based on a Geographic Information System (GIS) (Figure 3). The GIS uses spatial data for soil parameters, topography, field borders, and field length in relation to wind direction, wind barriers, and the general land use. On the basis of these data, a soil erodibility class, according to the dominant soil type, is derived for each particular field site.

Each soil erodibility class is assigned to a tolerable/critical field length (as discussed in the introductory chapter). The higher the erodibility of a particular soil, the shorter is the tolerable field length. This means, for example, that the tolerable length of a field dominated by loamy sands may exceed the length in a fine sandy field site by 150 m. Wind barriers enlarge the tolerable field length in proportion to their height. Since the tolerable field length changes with changing wind direction, the erosion risk is calculated for each direction separately. Thus, the method locates the fields that are at jeopardy for each wind direction. Fields are said to be potentially at risk from wind erosion when the actual field length exceeds the tolerable length. The GIS-based method provides a tool for the relevant authorities to simulate the effects of new plantings of wind barriers, changing field sizes or the consolidation of farming.

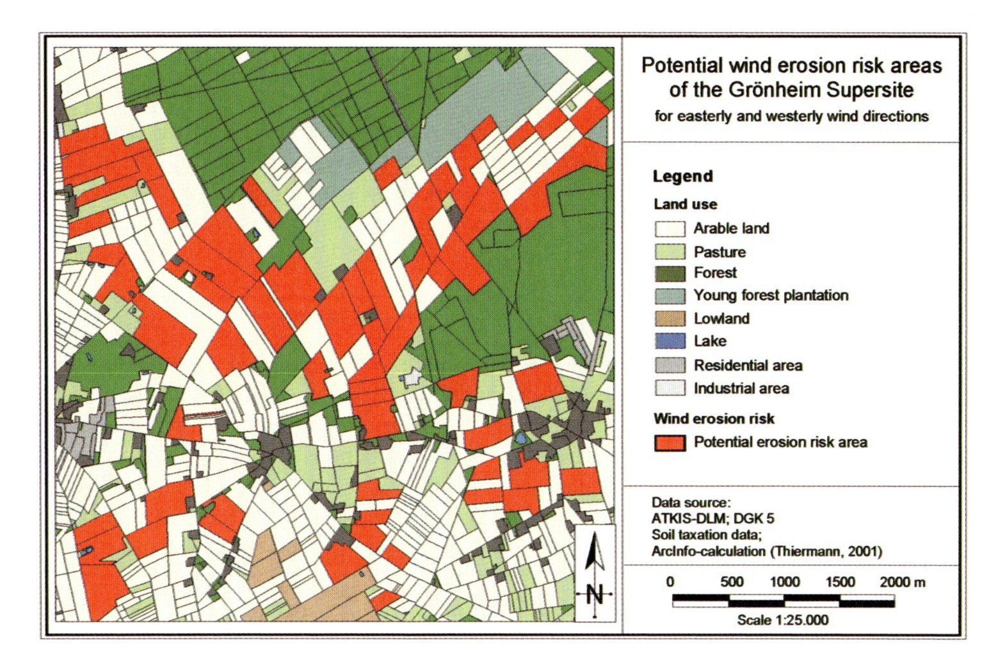

Figure 3. Estimation of the fields that are potentially at risk from wind erosion in the WEELS study area in Lower Saxony, Germany. Wind erosion risks are determined by means of a method based on a Geographic Information System (GIS).

Currently, an additional important application of this method is in relation to the German Soil Protection Law (Schäfer, 2001, personal communication). Soil quality standards will shortly be implemented to put this law into general practice. The implementation of the new standards will require, among other things, the assessment of the erosion risk of particular fields and the evaluation of available precautionary measures. At present it is proposed to use the GIS-method for this purpose.

Approaches to modelling the interactions between the factors that influence wind erosion. The complex interactions between the variables that control wind erosion and their high spatial and temporal variability are not properly accounted for in the procedure described above for determination of potentially erodible land. For more refined requirements such as estimating the effect of erosion on the long-term on yield, determining physical damage to plants, or calculating on-site and off-site economic costs of erosion, etc. better tools are needed. What is needed is an advanced, spatially distributed model that is capable of accommodating various time scales, taking into account different crop rotation periods, and able to simulate different management and climatic scenarios at the scale of anything up to and even beyond decades.

The requirements led, in the United States, to the development of an empirically based wind-erosion equation (WEQ) in the 1960s (Woodruff and Siddoway, 1965). The WEQ was used to predict potential average annual soil loss by determining the influence of several primary variables. However, in parallel with developments in the water-erosion field, these early models were found to be too empirical and had limited transferability to areas beyond their development (in Kansas). There were moves to develop models that were more physically – based and which allowed for the complex interactions between erosion and its controlling variables.

At present there are three main process-based wind erosion models (WEPS, WEAM, RWEQ). Most of them require detailed input data, which are not always available, and are not easily adapted to conditions or climates different from those in which they have been developed. In order to test a model that might be more applicable in the European context the EU Project, 'Wind Erosion on European Light Soils' (WEELS) developed a new model that is described below.'

The WEELS Wind Erosion Model. The WEELS model was developed with the objective of providing a suitable, GIS-based wind erosion model that is able to meet the primary needs of potential end-users. It was developed with the purpose of providing tools to model wind erosion events for a period of 30 years, to make predictions based on climatic change scenarios and on different land use scenarios. A further step would be to simulate average soil losses for different crop rotations. It was decided to test it with two sets of data. First it was tested against local knowledge of significant erosion events. Second it was tested against soil erosion estimated over a 30-year period using the ^{137}Cs method. It was also a primary requirement that the model should be able to work on already available data and be widely applicable within the EU.

These demands required partitioning into various submodules:

The 'WIND', 'WIND EROSIVITY' and the 'SOIL MOISTURE' submodules combine the factors that contribute to the temporal variations of climatic erosivity, while the 'SOIL ERODIBILITY', 'SOIL ROUGHNESS' and the 'LAND USE' modules predict the temporal soil and vegetative cover variables that control soil erodibility.

- WIND: In the first "analysis" step, the existing WAsP-Model analyses the regional wind climate by considering orographic conditions and the land use situation (roughness length as related to land use). The input data are time series (resolution 1 h.) of wind velocity and wind direction of regional meteorological stations. Based on the results of the regional analyses, in the "application" step, the WAsP-Model simulates the distribution of wind speed and direction at the test site with respect to the local topographic situation. The output consists of a grid map (25 25 m) of Weibull-A- and Weibull-K-parameters, calculated for 10 m above ground.

- WIND EROSIVITY: To estimate the erosivity of wind climates at the test site, the friction velocity is approximated, considering the surface roughness and the influence of wind barriers and hedges. For each grid of the zone, protected by a wind barrier, the friction velocity (u_*) of the wind is reduced for a given factor. The submodule output provides hourly time series for u_* for each grid point.

- SOIL MOISTURE: The water content of the uppermost 0-1 cm of the soil is computed by means of a simplified soil water balance equation, on a daily basis. Two soil moisture conditions (wet/dry) are defined (this is permissible in sandy soils). The soil moisture condition 'dry' means that the water content of the uppermost soil layer is below an empirically determined, critical soil water content. Only when the water content falls below this threshold is sediment transport/erosion possible. The output of this submodule provides days with wet/dry soil conditions for each grid point.

- SOIL ERODIBILITY: Soil erodibility is expressed as a dimensionless value, the so-called 'K-factor'. On the basis of a regression equation, empirically derived from wind tunnel experiments, this factor was calculated from data on soil texture and organic matter content of the topsoil. By means of further empirically determined 'transfer functions' values of u_{*t} (the threshold shear velocity at which transport begins), the relative sediment transport, and the maximum erodible amount of soil was derived from the K-Factor.

- SOIL ROUGHNESS: Aerodynamic roughness lengths (z_0) were determined from wind-tunnel experiments and data from the literature for the most important roughness conditions and cultivation measures. The annual variation of the crop cover between the sowing and the harvest was modelled by means of a phenology model. The phenology functions predict the development of the crop cover and the associated alteration of the surface roughness for the 8 prevalent crop types. A transfer function also allows the derivation of z_0 values from the land use type in combination with the K-factor.

- LAND USE: land use data were collected where possible from existing maps and farmers' records, but were also modelled where these were not available. The model for land use depended on average agricultural statistics for the communities and knowledge of rotations. Phenological data (from published sources) were used to convert these land use data into roughness values.

Preliminary Results of Simulations with the WEELS model. In its current state of development, the WEELS model outputs are hourly assessments of mean wind speeds (10 m above ground) and friction velocities as well as daily assessments of crop cover, associated surface (or tillage) roughness and top soil moisture as the main determinants for erosion processes. The actual erosion risk is given hourly in terms of the duration of erosive conditions and the corresponding maximum sediment transport rate, calculated with and without consideration of topsoil moisture. A simplified daily erosion/accumulation balance detects the impact of wind erosion in terms of soil loss and deposition rates.

As an example, the spatial distribution of erosion hours and erosion/accumulation balances is given below for wind erosion events at the Grönheim and Barnham test sites (Figures 4–7). The event of the 13th and 14th of March 1994 on parts of the Barnham site, which was recorded on video by the farmer. For these days the model showed that there was little erosion on fields covered by winter serials, cover crops and oil seed rape, but showed a total maximum duration within the two days of more than 8 erosion hours on fields that were not protected by vegetation (in that year these were sown to sugar beet, spring serials and maize). The maximum total soil loss an these fields was calculated at more than 7 t ha^{-1} in some places, particularly those not protected by hedges or other large obstacles from the westerly winds.

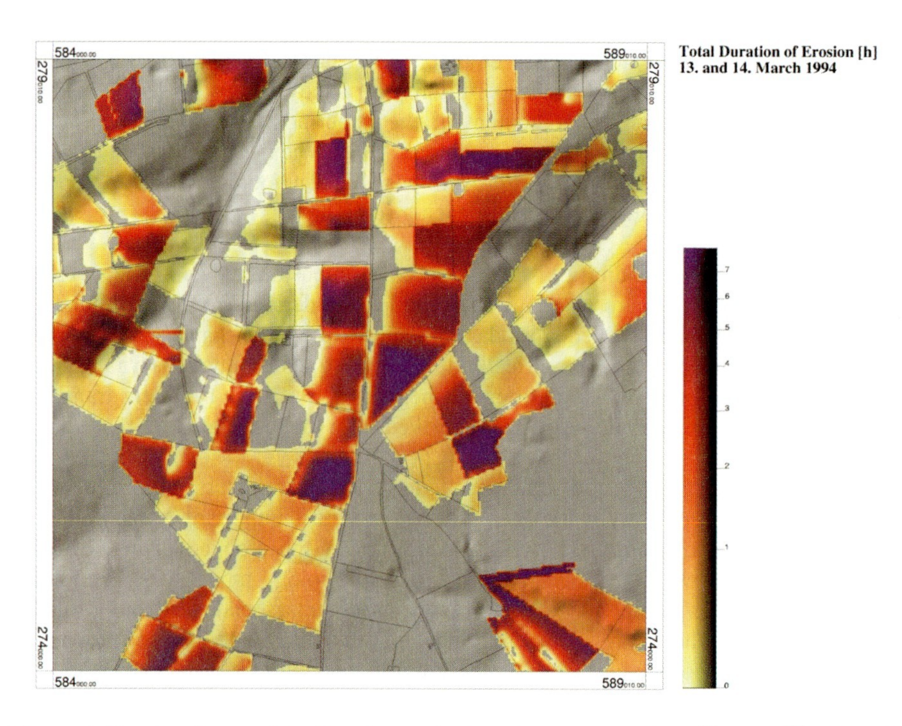

Figure 4: Erosion hours of the Barnham wind erosion event of the 13th and 14th of March 1994.

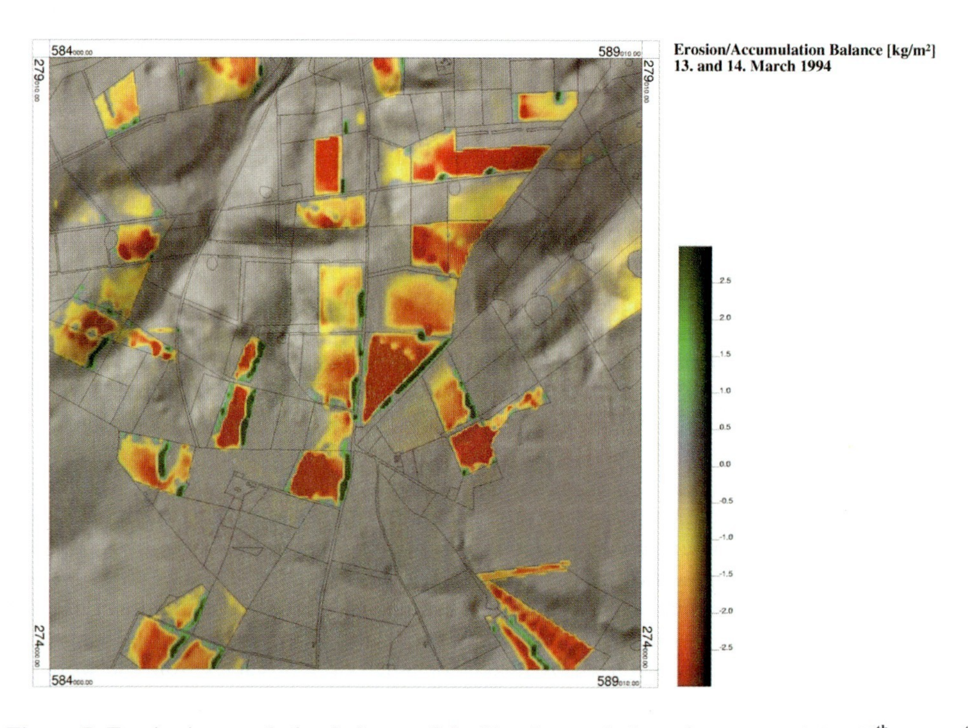

Figure 5: Erosion/accumulation balance of the Barnham wind erosion event of the 13th and 14th of March 1994.

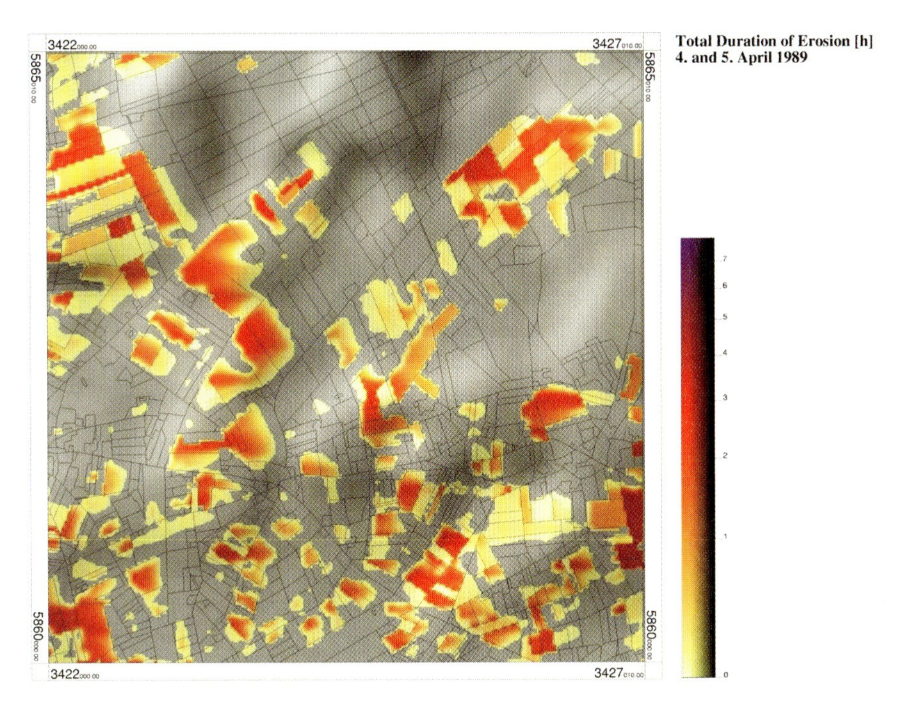

Figure 6: Erosion hours of the Grönheim wind erosion event of the 4th and 5th of April 1989.

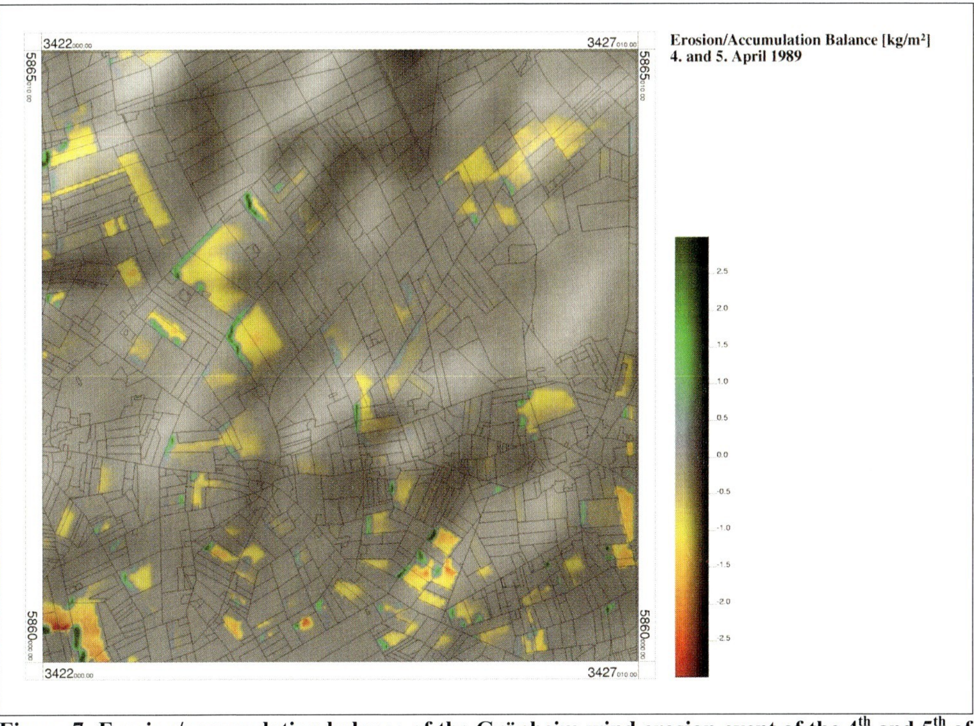

Figure 7: Erosion/accumulation balance of the Grönheim wind erosion event of the 4th and 5th of April 1989.

While the event at the Barnham site was the result partly of winds (that) hat reached 16m s^{-1} on the 13th of March (14.00 - 15.00 GMT), the easterlies of the 4th and 5th of April in 1989, which had maximum means of 14 m s^{-1} (4th April, 14.00 – 15.00 GMT) at the Grönheim test site caused less on-site damage. The maximum calculated soil loss was 5 t ha^{-1} and the corresponding maximum duration of erosive conditions was not more than 4 hours.

As the topsoil was simulated as being dry for both of these events, their comparison underlines the role of different wind directions in producing damage. The westerly winds at the Barnham site produced net accumulation on the eastern parts of the affected fields, especially in the hedges, while the corresponding pattern at the Grönheim sit was net accumulation to the west of the fields. Indeed, both at the Barnham site and at the WEELS Swedish site, there are accumulations in field boundaries, which are about 1 m high. At least one of these is known to have grown in the last 20 years.

In general, the model simulations of 28 years for the Barnham site (1970–1998) and 13 years (1981-1993) for the Grönheim site detected a higher erosion risk for the English test site, where the total mean soil loss was estimated at 1.56 t ha^{-1} y^{-1} and mean maximum soil loss of about 15.5 t ha^{-1} y^{-1}. The highest values were more than 3 t ha^{-1} in March, September and November. The total mean soil loss was less, at 0.43 t ha^{-1} y^{-1}, on the northern German test site. March and April experience highest erosion rates when they can exceed 2.5 t ha^{-1}. The total mean maximum soil loss at this site, of about 10.0 t ha^{-1} corresponds to a loss of about 0.65 mm.

Temporal variation

Introduction. All of the three general conditions that are necessary for wind erosion to take place – strong winds, a susceptible soil surface and no protective cover – are highly variable in time. Their temporal variation depends in turn on a large number of underlying variables.

Climate is the main forcing factor for temporal variation. The potential erodibility (inherent in the soil) is generally constant within a season and farming activities produce variations that are spatial rather than temporal. Climate determines when the soil surface is dry, in which condition point even brief gusts of strong wind will mobilise soil particles. It also determines when the wind speed is high enough to mobilise soil particles, which is usually 6 to 8 m s^{-1} (measured at 10 m above the ground).

Weather situations commonly associated with strong enough gusts are cyclones and weather fronts, but because these types of weather are often associated with precipitation, they seldom cause severe erosion. If there has not been a prolonged dry period, severe wind erosion generally occurs only after an extended period of strong winds. This is first because strong winds are those above the threshold speed; and second because drying capacity increases with wind speed.

The 'wind erosion season' is determined first by the phenology of the crop, for this controls when the soil is protected. Second, it is determined by when the soil surface is not covered by snow, or frozen, or too wet – requirements that are climatologically determined.

All this means that the main season for wind erosion begins in the spring when bare soil begins to dry and when cultivation loosens the surface. The season ends when a dense enough crop protects the soil. Thus, the main *wind erosion season* is usually from March to May and sometimes June (in Sweden). Soil erosion may occasionally take place before this main season, if strong winds and non-freezing conditions follow a dry period that is long enough to dry out the soil surface and destroy any crust; and after the harvest when the soil is again bare and if it is dry. Rain in the autumn and freeze/thaw cycles in the winter generally make the soil less susceptible.

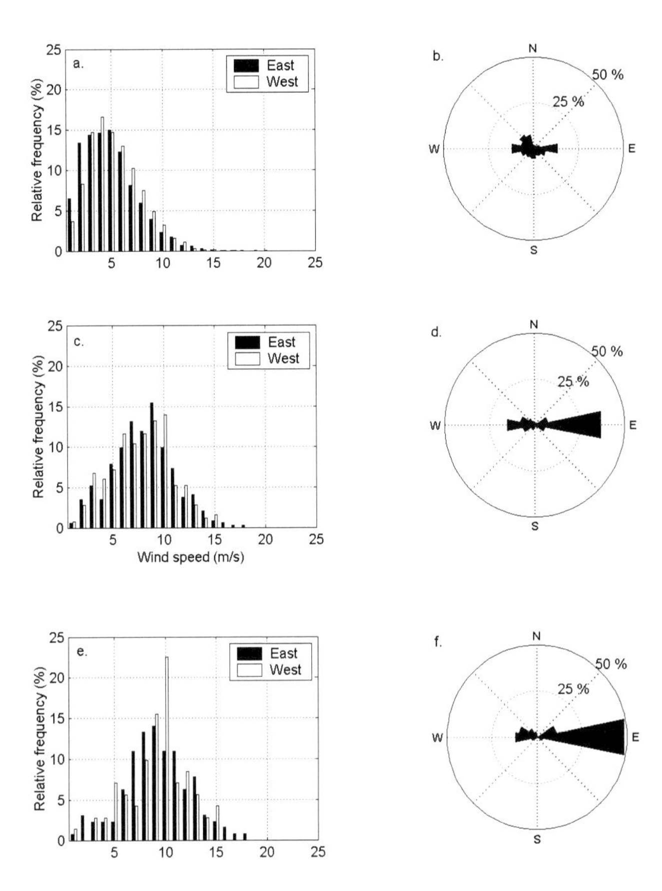

Figure 8. Wind climate ... to 1991) at the Sturup Ai... in southern Sweden. Left are the frequency distribution of wind speed (m s^{-1}) for easterly and westerly winds, and right are the distribution of wind directions in 16 sectors. Top row shows the conditions during the whole erosion season (locally adapted for sugar beet cultivation: April 15 to July 15). Middle row presents the situation all during identified erosion days, and the bottom row includes only those erosion days that required re-sowing of an area >32 ha (the 75 % percentile). The wind data are averages for 10 minutes. From Ekström *et al.* (2002).

Within the erosion season the occurrence of erosion depends on two factors: the farmer's activities and the weather. Tilling, seedbed preparation or sowing change the micro-relief and aggregate structure of the surface, and the application of manure, mulch or fertilizers influence the state of aggregation of the soil. A dense mulch may provide temporary protection. Irrigation can be regarded as equivalent to rainfall.

Weather situations associated with wind erosion. The following discussion relates mostly to Scania, in Sweden, where the analysis of weather patterns and wind erosion is most advances. It could quite easily be extended to other areas.

It has been known for some time that most of the severe wind erosion events in Scania are associated with strong easterly winds (Mattsson, 1987; Jönsson, 1992). This can be shown by using the recorded acreage of sugar beet that had enough crop damage by wind erosion to require re-sowing as an indicator for significant wind erosion. These data can be compared with the climatic forcing factors, which can be taken as days with less than a daily precipitation threshold of 3 mm and wind speed of over 9 m s^{-1}, which combined identify days when wind erosion is most likely ("erosion days"). Figure 8 shows that erosion damage increases with the mean wind speed and that significant wind erosion episodes are associated with two particular wind directions. These in turn are related to weather systems that are mainly governed by the large-scale distribution of atmospheric pressure (Ekström *et al.* 2002). One particular type of weather system accounts for 33%, and six for 83% of all the identified erosion days. The pressure pattern associated with these weather systems, which can be called *erosive pressure patterns*, are shown in Figure 9. Three of these erosive pressure patterns as associated with southerly to easterly winds, namely C7, C13, C9 (Figure 9a, e and f). Two, C5 and C1 (Figure 9b and d) are clearly associated with westerly to northerly winds.

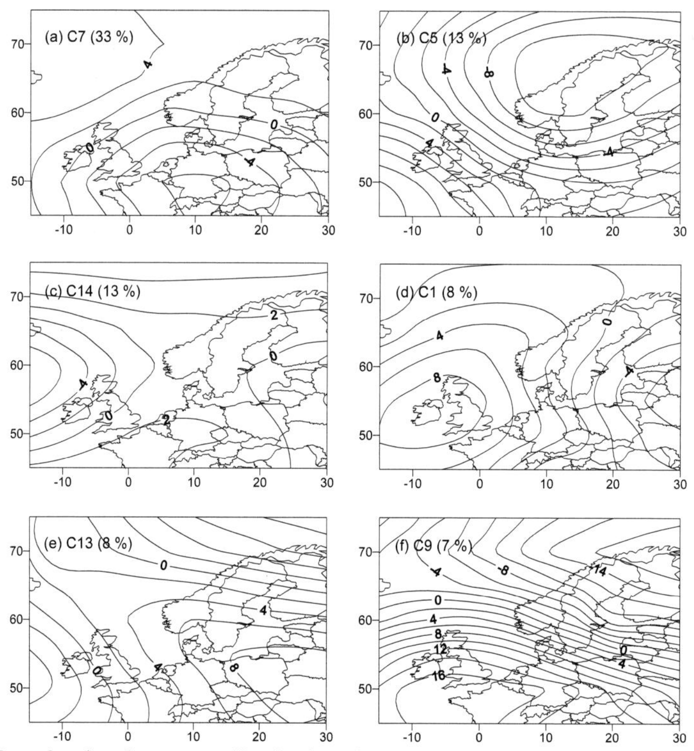

Figure 9. Maps showing the average distribution of atmospheric pressure (deviation from the mean) for each of the six erosive pressure patterns. The percentage of the identified erosion days associated with each pressure pattern is given. The closer the isobars (pressure isolines with units in hPa) are the higher the wind speed becomes. From Ekström et al. (2002).

One fundamental difference between weather systems that bring easterly and those that bring westerly winds is important to wind erosion. The westerly winds bring cyclones and frontal rain, but the easterly winds are associated with pressure patterns that block the rain-bearing weather systems. This means that in the easterly winds the soil surface dries out and becomes vulnerable to wind erosion. In westerly situations, cyclones pass over the area within about a day and are sometimes followed by another cyclone within a few days. The wind maximum associated with a cyclone normally occurs after the cold front, that is, immediately after the frontal system with its rain has passed. There is little chance for the soil to dry out, unless the fronts are weak and have little rainfall.

These results are for Scania, but the underlying atmospheric processes are the same for most of the area of northwestern Europe where wind erosion is a problem (as described above).

Climatic changes. Over the last few centuries there are several of periods of increased wind erosion activity. The question that arises is whether this was because of changes in the wind climate or changes agriculture. This question is important in the context of future agricultural policies in a situation where there may be climate change.

In the last decade there has been intense discussion regarding the evidence for changes to the storm climate of the North Atlantic and adjacent European regions and the prospect of a future change due to anthropogenic climate change. In a comprehensive review, the WASA Group (1998) concluded that there was no consistent observational evidence for a changing storm frequency and in the North Atlantic, but they also pointed out several shortcomings in the observational records. The Lund series (Figure 10; Bärring, 1999) is the first very long-term quality-controlled high-resolution air pressure series became available. It runs from 1780. Pressure data for one point alone cannot, of course, be used to gain insight regarding possible changes to the spatial patterns relevant for wind erosion, but these patterns are not independent of cyclonic activity. Figure 10 shows no long-term trend in the monthly 5 and 95 percentiles of pressure tendency. This indicates that there have been no long-term changes in cyclonic activity and, indirectly, no substantial changes to the distribution and frequency of specific pressure patterns.

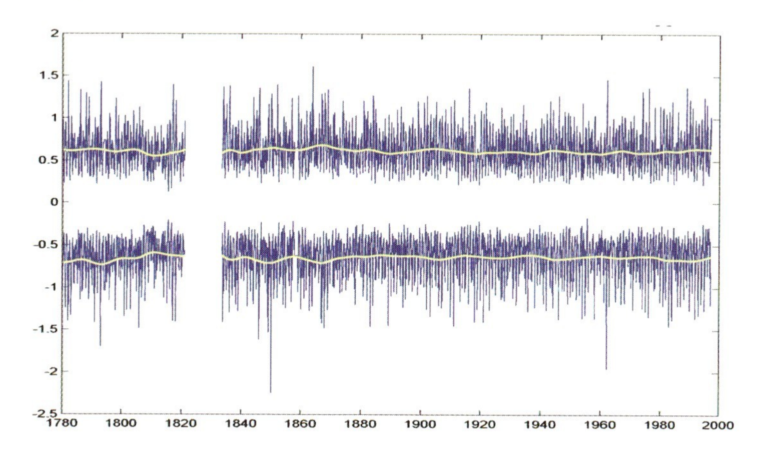

Figure 10. Variations in the five percentile (top) and 95 percentile (bottom) of pressure tendency (that is' pressure change per unit time) at Lund. The blue curves are the full monthly time series and the yellow lines are smoother series (from Bärring 1999).

So what of future climate change ? There have be a number of modelling studies of the storm and cyclonic activity in the North Atlantic and European region, but a consistent picture has not yet emerged (IPCC, 2001). Andersson (2001) evaluated the wind and pressure data produced by a Swedish dynamical regional downscaling model, RCA (Rummukainen *et al.*, 1998), but found that the results were much more sensitive to which of the two models were used than to simulations of changed climate. Thus, as yet, we can say nothing sure about what will happen to the wind climate, and so to wind erosivity in the future.

Conclusions

The last few years have seen advances in the understanding the distribution of wind erosion in Europe, and in particular in northwestern Europe. They have been made possible by better modelling and GIS technology. We now have a means of modelling and measuring wind erosion in areas of the order of 5 x 5 km. They have shown that the rate of erosion can be as high as in many areas affected by water erosion, and that we can reproduce the pattern of the process by modelling. Nonetheless, we are left with a gap between the modelled and the measured rates, and moreover the methods need very large amounts of data, to the point where they are unlikely to be able to be used, as they exists in many areas. We must rely on other methods to estimate the extent of the problem for planning and mitigation. These include gross estimates at the continental scale, which appear to many to be very crude as they stand, and some GIS-based methods for estimating the problem at the regional and local scales, which show more promise. They may lack the precision of the modelling and measuring methods we have just referred to, but they are probably adequate for planning and control.

As to the timing of wind erosion, we have also made advances, at least in some regions, where it appears that wind erosion events might be able to be predicted with some accuracy. The techniques of climate analysis used in this work could readily be applied to other regions. As the evidence now stands, it looks as if periods in which wind erosion was frequent in the past (some of which are mentioned in the introduction to this booklet) were probably more the result of changes in land use than of changes in climate, but the conclusion is not yet near certainty. When it comes to predicting the likelihood of wind erosion under different climate change scenarios, however, we are a long way from certitude.

References

Andersson, A. 2001. *The wind climate of northwestern Europe in SWECLIM regional climate scenarios*, *M.Sc. Thesis*, **80**, Department of Physical Geography, Lund University, 43 pp.

Bärring, L. 1999. Preliminary analysis of pressure tendency variability in the Lund record (1780–), southern Sweden, *International Conference on Climate Change and Variability – Past, Present and Future*, Tokyo, Japan, 13–17 September.

Bärring, L., Jönsson, P., Achberger, C., Ekström, M. and Alexandersson, H. 1999. The Lund record of meteorological instrument observations: monthly pressure 1780–1997, *International Journal of Climatology*, **19**, 1427-1443.

Beinhauer, R. and Kruse, B. 1994. Soil erosivity by wind in moderate climates, *Ecological Modelling*, **75/76**, 279-287.

Chepil, W.S. 1945. Dynamics of wind erosion: III. The transport capacity of the wind, *Soil Science*, **60** (4), 475–480.

Düwel, O., Schäfer, W. and Kuntze, H. 1994. The effect of soil surface roughness on soil transport by wind, in *Proceedings of the International Symposium on Wind Erosion in West Africa: the Problem and its Control*, Eds. Buerkert, B., Allison, B.E. and von Oppen, M., University of Hohenheim.

Ekström, M., Jönsson, P: and Bärring, L. 2002. in press. Climate Research. Synoptic patterns associated with wind erosion in southern Sweden (1973–1991).

European Environment Agency 1998. *Europe's Environment: The Second Assessment. European Environment Agency*. Elsevier, UK, 293 pp.

European Environment Agency 2000. *Down to earth: Soil degradation and sustainable development in Europe – A challenge for the 21st century, Environmental Issue Series,* **16**, Copenhagen 2000.

Fryrear, D.W. and Saleh, A. 1996. Wind erosion: Field length, *Soil Science*, **161** (6), 398–404.

IPCC, 2001: *IPCC, 2001.* Technical summary, in *Climate change 2001,* Cambridge University Press, Cambridge.

Jönsson, P. 1992. Wind erosion on sugar beet fields in Scania, southern Sweden, *Agricultural and Forest Meteorology*, **62**, 141-157.

Lyles, L. 1988. Basic wind erosion processes. *Agriculture, Ecosystems and Environment,* **22/23**, 91–101.

Mattsson, J.O. 1987. Wind erosion and climatic changes. Comments on the ecological crisis of Skåne during the 18th century. *Swedish Geographical Yearbook*, **63**, 94–108. (In Swedish, English summary).

Oldeman, L.R., Hakkeling, R.T.A. and Sombroek, W.G. 1991. *World map of human-induced soil degradation: an explanatory note, Global Assessment of Soil Degradation (GLASOD)*, ISRIC and United Nations Environment Programme (UNEP) / Winand Staring Centre-ISSS / Food and Agriculture Organization of the United Nations (FAO)-ITC, 27 pp.

Riksen, M.J.P. and de Graaff, J. 1999. *On-site and off-site effects of wind erosion on European light soils*. Wageningen, Erosion and Soil & Water Conservation Group (mimeographed).

Rummukainen, M., Räisänen, J., Ullerstig, A., Bringfelt, B., Hansson, U., Graham, P. and Willén, U. 1998. RCA-Rossby Centre regional atmospheric climate model: model description and results from the first multi-year simulation. *SMHI Reports Meteorology and Climatology,* **83**, 76 pp.

Sterk, G., López, M.V. and Arrúe, J.L. 1999. Saltation transport on a silt loam in Northeast Spain, *Land Degradation and. Development*, **10** (6), 545-554.

Thiermann, A., Sbresny, J. and Schäfer, W. 2000. Ermittlung der Erosionsgefährdung durch Wind, *Mitteilungen der Deutschen Bodenkundlichen Gesellschaft*, **92**, 104-107.

Van Lynden, G.W.J. 1995. European soil resources. Current status of soil degradation, causes impacts and need for action, *Nature and Environment,* **71**, Council of Europe Press, Strasburg.

Veen, P.H., Kampf, H. and Liro, A. 1997. *Nature development on former state farms in Poland,* Report within the framework of the Memorandum of Understanding for Nature Conservation between the Polish Ministry of Environmental Protection, Natural Resources and Forestry and the Dutch Ministry of Agriculture, Nature Management and Fisheries, 48 pp.

WASA Group 1998. Changing waves and storms in the Northeast Atlantic? *Bulletin of the American Meteorological Society,* **79**, 741-760.

Wilson, S.J. and Cooke, R.U. 1980. Wind erosion, in *Soil Erosion*, Eds. Kirkby, M.J. and Morgan, R.P.C., John Wiley and Sons, Chichester, 217-252.

Woodruff, N.P. and Siddoway, F.H. 1965. A wind erosion equation, *Proceedings of the Soil Science Society of America*, **29** (5), 602-608.

3. On-Site and Off-Site Effects of Wind Erosion

Dirk Goossens, Wageningen University

Introduction

The effects of wind erosion are either on-site or off-site (Table 1).

- On-site effects are those caused by wind erosion at or very close to the site where erosion occurs. The scale is from several metres to several hundreds of metres at maximum. The affected area is usually limited to the wind-eroded agricultural field itself, and/or the adjacent fields.
- Off-site effects include any others, initially generated by wind erosion, but at larger distances. They also include indirect effects, for example, damage caused by the deposition of the eroded particles, sometimes on very distant sites.

The relative significance of these two categories in any one area, and of their components, depends on two factors: the local soil type and the climate; and the prevailing agricultural system including the crops, and the duration and intensity of agricultural tillage procedures.

These influences differ markedly from region to region, and so too, therefore, does the relative importance of on-site and off-site wind effects, and their components. This overview summarises the effects of wind erosion in general terms, i.e., without geographic restrictions.

On-site effects

The effects of wind erosion on-site are either

long-term as in the progressive degradation of the topsoil; or
short- to medium-term, as in its effects on the current crop (or other vegetation).

Removal of fine particles. Soil particles of different size are not equally vulnerable to wind erosion. In general, small grains are easier to erode than coarse grains, but very fine grains cohere together more strongly than larger grains, making them resistant to wind erosion. There is thus an optimum grain size, at around 80-100 μm, where the particles are most susceptible to wind erosion (Bagnold, 1941). During erosion, these particle sizes are removed (provided they are available). Thus the general effect of wind erosion is a gradual coarsening of the topsoil (formation of an upper layer with coarse sand, concentration of the rock fragments on the surface). This, itself, leads to a serious degradation for several reasons:

- Soil nutrients are largely held by the fine particles. Clay particles also help to form aggregates, and these are important in controlling further wind erosion;

- The water economy in the topsoil degrades. A coarse sandy topsoil dries quickly, and this not only affects crops, but also increases the soil's vulnerability to subsequent wind erosion;

- A lower silt and clay content negatively enhances the eluviation (downwashing) of humus and other soil constituents into the soil, and it promotes acidification;

- To remain fertile, degraded sandy topsoil requires a significant amount of manure, which is much less the case for topsoil rich in silt and clay.

In special cases, wind erosion may even remove the entire surface soil (Chepil 1957a, 1957b; Zingg 1954), leaving behind sterile bedrock.

Table 1. Some on-site and off-site effects of wind erosion

On-site Effects	Off-site Effects
Soil degradation	*Short-term effects*
(1) Fine material may be removed by sorting, leaving a coarse lag	(1) Reduced visibility, affecting traffic safety
(2) Evacuation of organic matter	(2) Deposition of sediment on roads, in ditches, hedges, etc.
(3) Evacuation of soil nutrients	(3) Deposition of dust in houses, on cars, washing, etc.
(4) Degrading water economy in the topsoil	(4) Penetration of dust in machinery
(5) Degrading soil structure	(5) Deposition of dust on agricultural and industrial crops, ruining their quality
(6) Stimulated acidification of the topsoil	
Abrasion damage	*Long-term effects*
(1) Direct abrasion of crop tissue, resulting in lower yields and lower quality	(1) Penetration of dust and its constituents in the lungs, causing lung diseases and other respiratory problems
(2) Infection of crops due to the penetration of pathogens	(2) Absorption of airborne particulates by plants and animals, leading to a general poisening of the food chain
(3) Stimulated dust emission due to sandblasting of the surface toplayer	(3) Deposition of heavy metals and other eroded chemical substances to the ground, infecting the soil
Other damage	(4) Contamination of surface and ground water via deposition of airborne particles
(1) Infection, with pathogens or soil constituants, of adjacent uncontaminated fields and crops	(5) Increased eutrophication of surface and ground water
(2) Accumulation of low-quality wind-blown deposits on the fields	(6) Infection of remote uncontaminated areas, transforming these into new potential sources
(3) Building of sand accumulations at the field borders, covering of drainage ditches	
(4) Burrial of plants	
(5) Loss of seeds and seedlings	

***Removal of organic matter*:** Organic matter is essential to soil fertility (Chepil and Woodruff, 1963). Since most of it is bonded to clay particles (Zenchelsky *et al.*, 1976), which are themselves usually attached to the fine sand and silt (or form small aggregates), most soil particles lost in wind erosion have much more organic matter than their parent soil. Wind-blown sediment often has 20% more organic matter (Daniel and Langham, 1936). Medium- and fine-textured soils are especially vulnerable to this type of degradation.

A small proportion of soil organic matter is in the form of vegetable residue. Because of its very low density-to-size ratio, this residue is very easily eroded by the wind. Its removal weakens the structure of the soil, and also precludes its transformation into humus, which could have been absorbed by the upper soil layer.

Removal of other soil constituents. Many other soil constituents are removed in wind erosion. Several of these are indispensable to the agricultural quality of the soil. Among these are N, P, K, and other nutrients. Analyses of wind-blown sediments have clearly demonstrated the higher concentrations of these elements in the eroded particles compared to the parent soil (Duncan and Moldenhauer 1968, Sterk *et al.* 1996). Although the highest concentrations are in the suspended fraction (dust), the fluxes of N, P and K transported in the saltation layer, travelling close to the ground, are normally higher because of the higher proportion of the total material that is transported in this way. The suspended fraction is more important with respect to the degradation of the soil, however, since saltation results only in the local redistribution of the nutrients (usually within the field itself), whereas the nutrients in suspension almost all leave the parent field, and this often results in a regional loss of nutrients (Sterk *et al.*, 1996).

Seeds and seedlings may also be blown from the fields during wind erosion. This is a direct source of economic loss, especially if resowing is necessary.

Abrasion. Abrasion caused by the physical impact of airborne particles, may result in serious damage to crops. Damage can even occur during low to moderate erosion, since impacts by only a few particles is sufficient to destroy parts of the crop skin. Moreover, only short periods of exposure are needed for damage to occur (Armbrust, 1968). Thus, erosion damage can occur even when soil losses are well below those that damage the soil itself (Kimberlin *et al.*, 1977).

Abrading particles affect crops in different ways. They may reduce the survival and growth of seedlings; they may depress the yield; they may lower the marketability of vegetable crops; they may increase the susceptibility of plants to certain types of stress, including disease; they may contribute to the transmission to some plant pathogens; and they may affect changes in the plant's metabolic processes even before there is visual evidence of damage (Armbrust 1972, 1982, 1984; Chaflin *et al.*, 1973; Hayes, 1965; Skidmore, 1982).

Coarse (saltating) grains cause the most direct damage. Indirect damage due to penetration into the plant of various substances (chemical elements, pathogens) mainly occurs via the fine fraction (dust), since these substances are predominantly linked to the fine particles. Direct abrasion damage due to impacting dust has, however, also been observed (Dietrich, 1977).

Apart from the crops, the soil itself may be subject to abrasion. Bombardment of the surface layer can be very intense during erosion, especially on sandy soil. The general effects of these impacts are the maintenance (in time) and extension (in space) of the saltation process (and, thus, of the wind erosion), and the generation of dust. The impact of particles is one of the major mechanisms for dust emission, as has been demonstrated by many experiments (e.g. Shao *et al.* 1993, Rice *et al.* 1996, Gillette and Chen 1999).

Infection of adjacent fields. Short-distance transport of matter eroded from a field can infect adjacent ones. The infecting material may originate from the topsoil of the eroding field, or from crops that are abraded. Various types of matter can be transported, such as chemical residues, pathogens, weed seeds and plant residues.

Accretion of wind-borne sediment. Where sand transport is intense, sand may accumulate, as at the field boundaries. In many wind-eroded agricultural fields a sand blanket several centimetres thick is observed after the storm on the leeward side of the field. The extent of such blankets, which may completely bury crops, can be large. More important is the poor quality of the accumulated sediment. It usually consists of almost pure sand, contains very little organic matter and other soil nutrients, and it is very acid. It may also contain weed seeds and other undesirable plant residues that have been eroded from upwind. At the field borders, sand accumulations may fill drainage ditches; or build sand dunes that must then be taken out of production.

Significance. All these effects contribute to a general and progressive degradation of the topsoil, resulting in decreased yield. This can be compensated for only by intense and sophisticated use of fertiliser, by adequate tillage of the land, and by using improved crop varieties. These techniques are well developed in northern Europe, and in many instances mask the degradation in soil productivity caused by erosion. In other regions, for example in several parts of the US, the effective loss of agricultural production due to soil degradation by wind erosion has been estimated at several per cent, at least (Lyles, 1975). No data are available for northern Europe, although the losses will probably be lower than those in the US. However, several northern European cases have been documented where soil degradation due to wind erosion has become of major concern. In southern Sweden, for example, arable fields have already been removed from production because of the substantial decline of the soil productivity over years caused by wind erosion (Jönsson, 1992). In several areas, the degree of degradation has been so dramatic that even grass or trees planted on these fields simply do not grow. Thus, in areas where there is no compensation via the use of fertilizers and/or adequate tillage techniques, degradation of soil due to wind erosion may become rapidly visible, even in northern Europe.

Off-site effects

General considerations. The off-site effects of wind erosion arise directly from the transport and deposition of eroded particles. The distance of transport is usually more than a few hundreds of metres, but can reach thousands of kilometres. Off-site effects are always the result of particles fine enough to be transported in suspension. The size range is normally from less than a micron to a few tens of microns at maximum. Figure 1 shows the potential transport distance as a function of particle size. For normal windstorms, the coefficient of turbulent exchange (K_M) is between 10^3 and 10^5 cm^2 s^{-1}, usually around 10^4 cm^2 s^{-1}. Suspended particles leaving agricultural fields during wind erosion events normally have a grain size below 60 μm. The effects can be classified as either

- *Short-term.* Many particles settle rather quickly, and the problems they cause are restricted to short periods, usually in the order of hours.
- *Long-term.* Particles smaller than 20 μm may remain suspended for a long time, from several days to several months, and the problems they cause are more or less permanent.

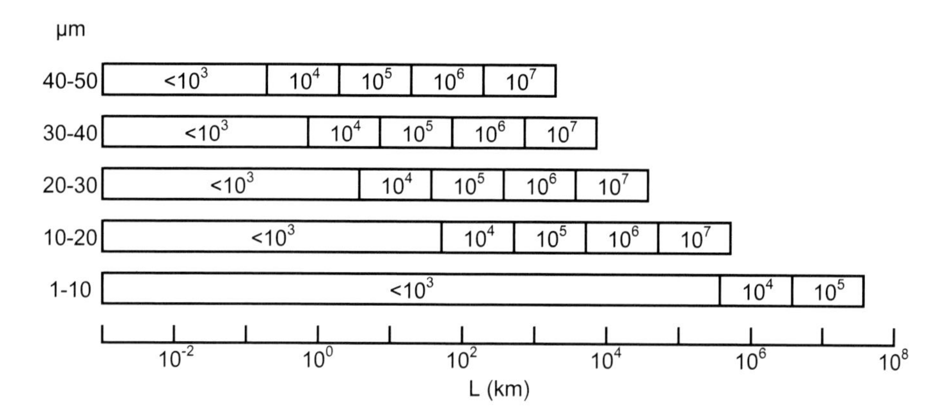

Fig. 1 The maximum distances likely to be travelled by different classes of quartz spheres, for K_M values between 10^3 and 10^7 cm^2s^{-1} (K_M = coefficient of turbulent exchange). The figure is for a wind velocity of 15 m s^{-1}. (After Tsoar and Pye, 1987)

Short-term effects. Where dust emission is intense, or occurs from large areas, blowing dust can reduce visibility, affecting, for example, traffic safety. This phenomenon occurs frequently in semi-arid areas with an open landscape, such as the Middle East, northern Africa, Australia, or parts of Asia and North America. It is less frequent in Europe, especially in northwestern Europe, where most fields are small. Likewise damage from the deposition of dust (or sand) on roads, or roadsides or drainage ditches, is less common in northwestern Europe although it has been reported in several countries (Robinson 1968, De Meester 1982, Poesen *et al.* 1996). The costs in the cleaning of the accumulations must be borne by public bodies (Duncan and Moldenhauer, 1968). In Seward County, Kansas, for example, the state highway department spent over $ 15 000 in 1996 to remove 965 tons of wind-blown sediment from only 150 m of highway and ditch (WERU website, 2001).

Dust can also cause considerable nuisance when it is deposited in houses, on cars, or when it penetrates into farm and other machinery. Again, these problems are not very severe in northwestern Europe, although in several rural and urban areas deposited soil dust has proved to be responsible for increasing the need for cleaning, maintenance and replacement, all of which incur expenditure (Stach and Podsiadlowski 1998).

In many semi-arid areas, deposited dust can directly affect the production of agricultural and industrial crops. A well-known example is cotton, which can become completely worthless, if dust penetrates the plants themselves. Poesen *et al.* (1996) reported a North-European example, where settling dust caused considerable damage to floriculture in northern Flanders. The hazards caused by the deposition (and accumulation) of wind-transported particles in northwestern Europe are much more important in the long-term.

Long-term effects. The dust leaving an eroded field carries organic matter, heavy metals, herbicides, pesticides, fertilisers, etc. Once deposited, these may contaminate surface and groundwater, sometimes eutrophying them. This problem is becoming increasingly important in northern Europe, where 1) large amounts of fertilisers, manure and pesticides are being used in agriculture, and 2) there is increasing contamination of soils by atmospheric pollutants such as heavy metals, dioxins, radionuclides, organic pollutants, xenobiotica, etc. The spread of pollutants carried by the dust from contaminated arable land has not been fully quantified, but must be considered important. Schulz (1992) proposed a value of 100 ng TEQ/kg for dioxin-contaminated arable land due to wind-induced emission of fine soil particles. This rate of emission is comparable to that from modern incineration plants. The quantities of deposited (and infiltrated) matter are very large, particularly because of the very large scale of the deposition area (millions of km^2). Infected areas may be very far from the sources, thus making control problematical. The gradual poisoning of several hundred km^2 in rural northeastern Belgium as a result of the deposition of particles emitted by the wind in heavily polluted upwind areas (Poesen *et al.*, 1996) is a good example. Another is the Aral Sea region, where very extended areas are currently becoming salinated because of the deposition of salt-enriched dust eroded from the desiccated sea basin (O'Hara *et al.* 2000). These types of contamination are especially important in the long-term, since they gradually change the composition of components of regional ecosystems.

Apart from problems of deposition and accumulation, eroded particles can also be harmful when they are still airborne. Particles smaller than 10 µm (the so-called PM_{10} fraction) are especially important since they are small enough to be taken into the body's respiratory system. They have been proved to be responsible for lung and other respiratory problems, in northern Europe, as elsewhere (Knottnerus 1985, van Nijf 1987, Pope *et al.* 1999). Pure quartz dust is most harmful in its crystalline form; the amorphous form is less dangerous. Various substances attached (or bound) to the particles, such as, for example, heavy metals, are even more dangerous. Their gradual accumulation in the human body can lead to serious illness. The problem is not restricted to humans: plants and animals, which may then be consumed by people, also absorb airborne particles. In this way aerosols directly contribute to the potential poisoning of the food chain.

Air quality has become a major issue in the US since the 1990s. The 1990 Federal Clean Air Act made the States responsible for monitoring and controlling the amount of PM_{10} particles. In the EU, guidelines have been published for various airborne components, among which are PM_{10} dust and total dust (see below).

Of course, besides wind and tillage erosion, many other sources contribute to the emission of airborne particulates, among which industry and traffic are the most important. However, in several wind erosion areas, the amounts of dust eroded from the land during a storm can be very considerable. It has even been suggested that, in some areas, wind erosion contributes more particulate matter to the atmosphere than all other sources combined (Kimberlin *et al.*, 1977). According to Fennelly (1976), natural soil dust constitutes almost half the total mass of particulate matter injected into the atmosphere worldwide. Thus, the impact of wind (and tillage) erosion in the atmospheric dust load should not be minimised, although its contribution strongly varies from region to region.

Significance. For northern Europe, long-term off-site effects of wind erosion are much more important than short-term off-site effects. Modern agriculture with its abundant use of fertilisers, manure, herbicides and pesticides, and its intensive cultivation of the land, is one major reason. Equally important are the climatic conditions, with relatively large amounts of rainfall, which on one hand reduce the episodes of wind erosion, but on the other hand promote the deposition of airborne particulates (wet deposition), and, very important, guarantee an efficient percolation of the harmful substances to the groundwater. The real significance of the off-site effects is that areas located very distant from the original source are becoming polluted. There is much transboundary import (and export) of environmental degradation in Europe, not least in the form of dust.

Policy considerations

Policy for wind erosion must first be based on the identification of vulnerable areas (Chapter 2), and then the development of measures to prevent, or at least control, the erosion (Chapter 4). It is difficult to control erosion on-site (in the fields), since the diminishing agricultural production it causes is very easily compensated with manure and fertilisers (although severe restrictions currently exist in many areas regarding their application). Nevertheless, the sustainability of agriculture, which is jeopardised by erosion, and which cannot be compensated in these ways, is also an important concern of on-site policy. Farmers should be supported to maintain sustainability. Soil conservation codes, as in the British *Code of Good Agricultural Practice for the Protection of Soil* (MAFF, 1998), or the German *Bundes-Bodenschutzgesetz* (1998), could be important stimulants if they were accompanied by financial support.

As to off-site effects, the most urgent need is to set up regulations that prevent, or at least diminish, the export of harmful substances to uninfected areas. Since law cannot forbid atmospheric transport, measures would have to focus on source areas. Some forms of policy regulation in this area are discussed in Chapter 4. There is, however, also an urgent need for more systematic verification and monitoring of the transport of harmful matter, as well as regulation of the import or export of contaminated soil. Measures should be taken to prevent the spread of infection, in particular because wind erosion may transform import areas into new sources of pollutants. The use of heavily polluted industrial soil to stabilise hundreds of kilometres of field roads in northeastern Belgium and the southeastern Netherlands, which are very susceptible to wind erosion and serve as important dust sources all year (Poesen *et al.* 1996), is an example that could have been prevented had adequate policy regulations existed.

As far as the medical aspects are concerned, European regulations exist with respect to several airborne substances such as SO_2, NO_2, Pb, fine particles (including soot and PM_{10}) and total suspended dust (EU Guidelines 1996/62/EG and 1999/30/EG). For other substances such as O_3, CO, C_6H_6, HCl, Cd, Ar, Ni and Hg, there are no European regulations, and although recommendations have been proposed for several of these products, they do not have the status of law. The same is true for several recommendations regarding the concentration of harmful substances in deposited (not airborne) dust. An important policy target should be to develop an official European jurisdiction with respect to these recommendations, giving them status in international European law. In addition, regulations should be elaborated for various other chemical elements present in atmospheric soil dust (not only, or predominantly, heavy metals).

Finally, more data are needed to document the current status of wind erosion in northern Europe. The process of wind erosion, and of soil particle emission due to agricultural tillage, has not yet been given the attention it requires. However, the data that are available indicate that these processes are among the primary factors that affect the quality of the north European ecosystems. Policy makers should consider these processes more directly than they have done, when they develop new regulations related to the environment.

References

Armbrust, D.V. 1968. Windblown soil abrasive injury to cotton plants, *Agronomy Journal*, **60** (6), 622-625.

Armbrust, D.V. 1972. Recovery and nutrient content of sand-blasted soybean seedlings, *Agronomy Journal*, **64** (6), 707-708.

Armbrust, D.V. 1982. Physiological responses to wind and sandblast by grain sorghum plants, *Agronomy Journal*, **74** (1), 133-135.

Armbrust, D.V. 1984. Wind and sandblast injury to field crops, *Agronomy Journal*, **76** (6), 991-993.

Bagnold, R.A. 1941. *The physics of blown sand and desert dunes*, Methuen, London, 256 pp.

Bundes-Bodenschutzgesetz 1998. *Gesetz zum Schutz vor schädlichen Bodenveränderungen und der Sanierung von Altlasten*. Bundes-Bodenschutzgesetz von 17 März 1998 (BGBI. I S. 502).

Chaflin, L.E., Stuteville, D.L. and Armbrust, D.V. 1973. Windblown soil in the epidemiology of bacterial leaf spot of alfalfa and common blight of beans. *Phytopathology*, **63**: 1417-1419.

Chepil, W.S. 1957a. Sedimentary characteristics of dust storms. I. Sorting of wind-eroded soil material, *American Journal of Science*, **255** (1), 12-22.

Chepil, W.S. 1957b. Sedimentary characteristics of dust storms. III. Composition of suspended dust, *American Journal of Science*, **255** (3), 206-213.

Chepil, W.S. and Woodruff, N.P. 1963. The physics of wind erosion and its control, *Advances in Agronomy*, **15**, 211-302.

Daniel, H.A. and Langham, W.H. 1936. The effect of wind erosion and cultivation on the total nitrogen and organic matter content of soils in the southern High Plains, *Journal of the American Society of Agronomy*, **28**, 587-596.

De Meester, S. 1982. *Onderzoek omtrent de deflatiegevoeligheid van lemig zand en lichte zandleem*, MSc. thesis, K.U. Leuven, 117 pp.

Dietrich, R.V. 1977. Impact abrasion of harder by softer materials, *Journal of Geology*, **85** (2), 242-246.

Duncan, E.R. and Moldenhauer, W.C. 1968. *Controlling Wind Erosion in Iowa*, Cooperative Extension Service, Iowa State University, Ames, Iowa, 6 pp.

Fennelly, P.F. 1976. The origin and influence of airborne particulates, *American Scientist*, **64,** 46-56.

Gillette, D.A. and Chen, W. 1999. Size distributions of saltating grains: an important variable in the production of suspended particles, *Earth Surface Processes and Landforms*, **24**, 449-462.

Hayes, W.A. 1965. Wind erosion equation useful in designing northeastern crop protection, *Journal of Soil and Water Conservation*, **20** (4), 153-155.

Jönsson, P. 1992. Wind erosion on sugar beet fields in Scania, southern Sweden. *Agricultural and Forest Meteorology*, **62**: 141-157.

Kimberlin, L.W., Hidelbaugh, A.L. and Grunewald, A.R. 1977. The potential wind erosion problem in the United States, *Transactions of the American Society of Agricultural Engineers*, **20** (3), 873-879.

Knottnerus, D.J.C. 1985. *Verstuiven van grond: Beschouwingen over te nemen maatregelen, rapportering van gedaan onderzoek*, Nota **144**, Instituut voor Bodemvruchtbaarheid, 57 pp.

Lyles, L. 1975. Possible effects of wind erosion on soil productivity, *Journal of Soil and Water Conservation*, **30** (6), 279-283.

MAFF 1998. *Code of Good Agricultural Practice for the Protection of Soil*, Ministry of Agriculture, Fisheries and Foods, Welsh Office Agriculture Department, MAFF Publications, London, 66 pp.

O'Hara, S.L., Wiggs, G.F.S., Mamedov, B.K., Davidson, G. and Hubbard, R.B. 2000. Exposure to airborne dust contaminated with organophosphate pesticides in the Aral Sea region, *The Lancet*, **355**, 627-628.

Poesen, J., Govers, G. and Goossens, D.1996. Verdichting en erosie van de bodem in Vlaanderen. *Tijdschrift Belg. Ver. Aardr. Studies (BEVAS)*, 1996-2, 141-181.

Pope, C.A., Hill, R.W., and Villegas, G.M. 1999. Particulate air pollution and daily mortality on Utah's Wasatch Front, *Environmental Health Perspectives*, **107** (7), 567-573.

Rice, M.A., Willetts, B.B. and McEwan, I.K. 1996. Wind erosion of crusted soil sediments, *Earth Surface Processes and Landforms*, **21** (3), 279-294.

Robinson, D.N. 1968. Soil erosion by wind in East Lincolnshire, March 1968, *East Midlands Geographer*, **4** (30), 351-362.

Schulz, D. 1992. Obergrenze für den Dioxingehalt von Ackerböden, *Zeitschrift der Umweltchemie und Ökotoxikologie*, **4** (4), 207-209.

Shao, Y., Raupach, M.R. and Findlater, P.A. 1993. Effect of saltation bombardment on the entrainment of dust by wind, *Journal of Geophysical Research*, **98** (D7), 12719-12726.

Skidmore, E.L. 1982. Soil and water management and conservation: wind erosion, in *Handbook of soils and climate in agriculture*, Ed. Kilmer, V.J., CRC Series in Agriculture, CRC Press, Boca Raton, 371-399.

Stach, A. and Podsiadlowski, S. 1998. *The effect of wind erosion on the spatial variability of cultivated soils in the Wielkopolska region (Poland)*, International Conference on Agricultural Engineering (AgEng), Oslo 1998, Paper No. 98-C-089.

Sterk, G., Hermann, L. and Bationo, A. 1996. Wind-blown nutrient transport and soil productivity change in southwest Niger, *Land Degradation and Development*, **7** (4), 325-335.

Tsoar, H. and Pye, K. 1987. Dust transport and the question of desert loess formation, *Sedimentology*, **34** (1), 139-153.

Van Nijf, A. 1987. *Bestrijding van winderosie in drie landbouwgebieden in Nederland. Inventarisatie en evaluatie*, Thesis HBCS Velp, Arnhem, 56 pp.

WERU website 2000. Wind Erosion Research Unit, Kansas State University, US. http://www.weru.ksu.edu/problem.html.

Zenchelsky, S.T., Delany, A.C. and Pickett, R.A. 1976. The organic component of wind-blown soil aerosol as a function of wind velocity, *Soil Science*, **122,** 129-132.

Zingg, A.W. 1954. The wind erosion problem in the Great Plains, *Transactions of the American Geophysical Union*, **35**, 252-258.

4. What to do about Wind Erosion

Michel Riksen, Floor Brouwer and Wim Spaan at Wageningen University and José Luis Arrúe and María Victoria López at Aula Dei Experimental Station, Zaragoza

Introduction

One of the first attempts to control wind erosion (or at least its results, in sand drifts) occurred in the Veluwe, in the Netherlands, as early as the 16th century (Schimmel, 1975). Until 1898, however, the result of these measures was limited, partly because of a lack of co-operation from farmers. The wind erosion problem on the Veluwe was finally controlled after 1898 by re-afforestation by the government. There are many examples of other attempts to control wind erosion from all over Europe.

In north-western Europe several changes in agricultural practice and in the re-allotment of land over the last few decades have increased the hazard of wind erosion on agricultural land. They include the intensification of production, increasing sizes of fields, the intense use of heavy machinery and the disappearance of hedges at field borders (Riksen and de Graaff, 2001). As the body of evidence of damage to crops and of off-site damage has grown, farmers and policymakers have started to pay more attention to wind erosion control measures.

In southern Europe the area affected by severe wind erosion appears to be fairly limited and here wind erosion control is best approached in the context of more general soil conservation policies and measures.

A variety of measures has now been applied on farms to control wind erosion, among them the application of land management practices (like tillage, crop rotation), physical control measures (such as windbreaks) and extra agronomic measures (for example the growth of catch crops, and the use of cover or nursing crops). Most farmers, however, only use measures to protect their main crops. In most contemporary cropping systems there is a lack of protection of the soil in the period between harvest and the establishment of the next crop.

Policy makers still pays little attention to the off-site and the long-term effects of wind erosion. This chapter forms the basis for a discussion of measures that might minimise the wind erosion problem in Europe. It aims to contribute to the development of a general European policy towards wind erosion, based on an overview of:

- The most commonly applied measures, their effect and their adoption by farmers;
- Present policies concerning the prevention of wind erosion.

Measures to minimise the risk of wind erosion

High production and low input costs are no longer the only objectives in farming policy. In the last few decades society has become more aware of environmental problems and the value of landscape. Farmers are now obliged to take care of the land they farm. They have acquired a new role as the managers of the countryside. This means that they need good farming practices to safeguard the means of production as well as the beauty of the countryside. In some EU States this has been translated into a Code of Good Agricultural Practice, as in Great Britain and Germany. However, because the social costs and benefits (those that accrue to society and not to private individuals) can be high in the case of wind erosion control, it is also a task of society as a whole to make it possible for farmers to farm according these codes and regulations.

In areas at risk of wind erosion the general objectives of Good Agricultural Practice should be:

- Sustainable use of the means of production by avoiding crop damage, by ensuring that the rate of erosion does not exceed that of soil formation, and by restoring fertility where it has been lost;
- Minimisation of off-site damage: public and private property and public health should be protected against the effects of sand and dust eroded from agricultural land.

"Good Agricultural Practice" needs to include measures to meet these objectives. Where wind erosion is concerned, this may mean revision of existing codes.

The measures used by farmers can be grouped by the effect they intend:

a ***Wind velocity reduction***: drilling plant rows perpendicular to the direction of the prevailing wind, growing cover crops and planting shelterbelts.
b ***Soil stabilisation and increase roughness***: conservation tillage, furrow pressing, marling.
c ***Soil protection:*** artificial crusts, plastic foils, mulching.
d ***Minimising the risk of crop failure***: avoiding vulnerable crops on the most exposed fields, drilling seed deep, changing land use.

a. Wind velocity reduction

The aim here is to reduce the wind velocity at the soil surface. It can be temporary or more permanent.

Strip cropping A field can be made shorter in the direction of the prevailing wind, permanently or temporarily' by strip cropping. This prevents the wind from reaching the threshold speed for erosion anywhere in the field. This measure is only successful if crops in the adjoining strips have different phenological characteristics, so that only a limited number of strips are susceptible to wind erosion throughout the year. Leaving unploughed strips of stubble has the same effect (Hudson, 1986). In Europe this method is not commonly used to prevent wind erosion, perhaps because farmers are not aware of the practice or because the method does not fit in their farm management.

Windbreaks. An alternative is to plant a windbreak (Figure 1). Despite many attempts to use this method, it has often only partially solved the problem. Hedges and shelterbelts give protection downwind for a distance of 10 or 12 times the height of the windbreak (Morgan, 1995), depending on shape, porosity and alignment with the wind direction. In practice most shelterbelts are planted along roads and protect a very small area of the fields. Moreover, the planting of shelterbelts is in most cases prohibitively expensive. Costs are between €1000 - €2000 per 100 m (£600 and £1300). There is also loss of productive land and an increase in the operation times and operation costs of machinery. Yet too little is known about the net benefit of windbreaks. Many studies show a positive effect on crop yield varying from 3% up to 50% (Nuberg, 1998), but also a yield reduction close to the belt of between 0% and 47%. The figures also show that the net effect is influenced by factors other than the reduction of wind erosion, for example shelterbelt network can also positively influence the local microclimate by reducing the drying effect of wind. The overall effect varies therefore not only with the erosion risk but also with other local parameters. Shelterbelts (or shelterbelt networks) also have a positive effect on reducing the off-site effects of wind erosion, because they trap sand and dust. They can help to reduce the spread of diseases, the sandblasting of crops on adjacent fields, sedimentation in ditches and on roads, and the accumulation of sand and dust in houses. New windbreaks should be planned and funded with all these issues in mind, particularly who pays and who benefits.

Figure 1: Two year old wind break damaged wildlife (photo M. Riksen)

To be effective, windbreaks require small fields, and it is well known that the size and shape of a field have a critical influence on the work rate of field machinery, through the effect on the amount of turning per unit area and therefore the amount of unproductive time. However studies have found that the economy-of-scale argument applies only to very small fields (at least for mechanisation), and only marginal benefits are gained from increasing the size of a field above 5 ha (Fry, 1994). The proportion of compacted headland is greater in smaller fields, and in ones that are squarer, so that the most efficient fields are long and rectangular and about 5 ha in size. However the ideal field size and shape also depends on the height, shape and orientation towards the wind direction of the shelterbelt.

The adoption of windbreaks by farmers is very variable. In Denmark they have been widely adopted by the farmers because of good co-operation between government, farmers and contractors. There is good financial and professional support. There is said to be successful planting of 900 km of shelterbelt per year in Denmark (Als, 1989). Attempts in other countries, however, have not equalled this success. There are many reasons, such as the lack of support from the government and non-co-operation by the farmers.

Crop rows at right angles to the prevailing wind direction. Aligning crop rows perpendicular on the prevailing wind direction creates a micro-relief that protects the surface in the period between field preparation and a sufficient vegetation cover (Zobeck, 1991). The increased effective soil roughness created by ridges with this orientation, shelters part of the soil surface and reduces the wind shear and thus erosion. The ridges slow down the saltation process and intercept moving particles. The sheltering effect of the ridges depends on their height and spacing. The size of the furrows created by the planting sugar beet and cereal crops is far lower than for potatoes, so that potato crops are generally less vulnerable to wind erosion. In practice, most farmers in Europe plough parallel to the longest field side, because this minimises the time of operation of their machinery and the area of the headland. They weigh these advantages against the hazard of erosion.

Catch crops. Catch crops are crops sown after the main crop has been harvested. This measure is often used in groundwater-protection zones to 'catch' the nitrogen remaining in the topsoil. However it also has a positive effect in reducing wind erosion. It produces a ground cover in the autumn and spring periods in which bare fields would be susceptible to wind erosion. The most commonly used catch crops in northwestern Europe are yellow mustard (*Sinapis alba*), *phacelia, Raphanus sativus* and winter rye. Most catch crops are killed off by winter frosts, and delay during late winter and early spring, leaving a good mulch. Winter rye is ploughed in or killed by a herbicide before drilling in spring. Potatoes are planted in the rye, which is killed just before the emergence of the potato plants (Eppink and Spaan, 1989; Riksen and de Graaff, 2001). Catch cropping has become a common practice, especially on mixed farms. In some areas it is forbidden to manure land in the period between harvest and the next crop in spring, unless a catch crop is grown. The problem with catch crops is that they need to fit into a crop rotation. This means that they are not always feasible, as after a potato or sugar beet crop, both of which are harvested late autumn or early winter.

Cover crops or nurse crops. Unlike catch crops, cover crops or nurse crops give protection until the main crop has reached the stage beyond which wind erosion risk is very small. The most commonly used cover crops are spring barley (Figure 2) and winter rye. The "Dutch Rye System" is to broadcast winter rye before the November of the autumn preceding the planting of potatoes or sugar beet. The rye is killed with herbicide if its protection is not needed before the sowing of the beet, while the potatoes are planted in the rye, which is killed just before the emergence of the potato plants (Eppink and Spaan, 1989). Barley as a cover crop is sown in springtime just before or together with the main crop. The barley emerges much more quickly than the other crop and protects them as they emerge. The barley is killed just before the development stage of the main crop. In general, this measure is seen as very effective by farmers, although on some occasions it can be difficult to kill off the cover crop, and this can lead to crop competition and the need to use extra herbicides. In the case of a nurse crop like spring barley, which is sown two or three weeks ahead the main crop, there is still a short period of high wind erosion risk. Under extreme circumstances the nurse crop does not give enough protection and both nurse crop and the main crop need to be re-drilled (Riksen and de Graaff, 2001).

Figure 2 Sugar beet sown between established rows of barley (Photo IACR-Broom's Barn, Highham England)

Inter-cropping. To minimise the period without vegetation in a crop rotation, two crops can be sown at the same time; after the harvesting of the main crop, the second crop providing good soil cover. An example of this system is the one used in Sweden, where farmers combine spring barley with grass. The barley and grass seeds are sown at the same depth, or the barley is sown at 2-3 cm depth and the grass at 1 cm depth. In both cases the grass is left as the crop for the next year, so that the costs for the extra sowing are not entirely an extra expense, although the re-sowing or sowing of clover in between the grass can be necessary. The sowing of barley and grass at the same depth is to prevent the grass seeds from blowing away (Borstlap, 1999).

b. Soil stabilisation and roughness increase

The aim, with these types of measure, is to increase the threshold wind-speed of erosion by making the soil particles less detachable.

Tillage to keep the soil rough. In regions prone to wind erosion it is advisable to leave the soil as rough as possible. Large soil aggregates are less susceptible to wind erosion and reduce the wind velocity at the surface. This measure is used in the Netherlands where the desinfection of potato fields, which lasts till late autumn, does not permit the use of a catch or a cover crop (Eppink and Spaan, 1989). After desinfection the surface is very smooth, because the soil aggregates have been destroyed, and thus the soil is extremely susceptible to blowing. The same holds for potato and sugar beet fields, harvested after October. In all these cases, the farmers keep the fields as rough as possible, throughout the winter, by tilling the land perpendicularly to the dominant erosive wind direction. Machines are available which allow both desinfection and the sowing of a winter crop.

Furrow pressing. If the topsoil contains enough silt and clay, the ploughing and pressing of the soil at the same time can form an erosion-resistant surface (Figure 3). Adequate moisture is needed if pressing is to provide a stable surface. This is a commonly used practice in North European countries (Riksen and de Graaff, 2001). Most farmers use this measure to improve the water holding capacity of the topsoil, as well as to protect the land from erosion. The main extra cost is the investment in the so-called Cambridge roller.

Figure 3 Field after plough-and-press and drilling at right angles to the prevailing wind (photo Broom's Barn)

Reduced tillage. On a field with light soils, the number of tillage operations is less than on other types of soil, and reducing these even further can have benefits in terms of preventing wind erosion. If, alternatively, the soil is harrowed before sowing, most of the aggregates are destroyed to leave a fine soil structure, which is prone to wind erosion. Reduced tillage also controls wind erosion by allowing weeds and crop residues to remain on the land. The weeds can be sprayed of with herbicide, approximately 10 days before sowing. Just before sowing the land is tilled with a cultivator to break up the weeds and create a seedbed. This measure requires less input in terms of cultivation, although more in terms of chemicals. Special sowing machines may be needed if the seed is sown directly into the killed weeds or crop residues (direct sowing).

Improving the organic matter content. The use of organic manure has diminished since the introduction of cheap fertilisers, and this has resulted in a dramatic decline of soil organic matter content. This, in turn, has increased the erodibility of light sandy soils. To avoid this problem, farmers in the cut-over peat soils in the Netherlands plough deep to mix the topsoil with peat from the subsoil. This has only a temporary effect because of oxidation, which reduces particle size of the organic matter to a point where it can easily be blown away. Raising the organic matter content of topsoil sufficiently to control erosion needs large quantities of manure or compost. Nonetheless, using organic manure, as from catch or green manure crops, can reduce the erosion risk. A green manure crop adds more of organic matter than any other method.

Marling. Increasing the clay content of the soil to 8% increases aggregate stability and minimises the erosion risk. On many European sandy soils, a clay layer near the soil surface was traditionally used to improve the land. In some regions, as in much of East Anglia, one can still see the clay pits on aerial photos (Prince, 1962). Today this measure is generally too expensive.

c. Soil protection

Some measures are aimed to protect the soil against erosive winds by applying different materials on the soil surface.

Liquid manure. Liquid manure or slurry can be applied as a thin layer (15 ton ha^{-1}). This forms a good protective layer, which can hold for about six weeks. Slurry from bovine animals gives the best results because it contains more adhesive substance than in slurry from other livestock. In the Netherlands this practice is only allowed in restricted areas with high wind erosion risk. Some farmers apply manure only on the locations they know to be prone to erode, and at times that it is likely.

Synthetic stabilisers. Synthetic stabilisers such as PVA (polyvinylacetate) emulsions or PAM (polyacrylamides) sprayed onto the soil surface after drilling can provide temporary protection for high-value crops. For lower-value crops this practice is too expensive. Some farmers have synthetic stabilisers in stock as a last-minute precaution. In The Netherlands four other commercial products have governmental permission for use in the prevention of wind erosion: cellulose from the paper industry, a product based on wheat starch and sodium bicarbonate, a product based on calcium-lignosulfonate, and a product based on magnesium-lignosulfate. These products are industrial by-products and less synthetic than PVA or PAM. They are somewhat more expensive than manure, but their performance is much better. However, to be able to use these products, relative large investments in silos for storage and special machinery for spreading are needed. Their high costs restricts the use of these products to horticulture rather than arable farming.

Straw cover and residues. Many farmers who till light sandy soils leave plant residues or stubble on the field and plough it in a few weeks before sowing the next crop. Crops like sugar beet and potato produce a small amount of crop residue which decomposes rapidly. In the Netherlands the use of straw cover was at one time very common (Eppink and Spaan, 1989), but, because of its expense it is no longer in use in arable crops like potato and sugar beet. Straw planting is presently only used in high value crops like flower bulbs.

d. Minimising the risk of crop failure

The main aim of these measures is to lower the financial risk of farming on highly erodible land.

Excluding vulnerable crops. Farmers can avoid financial risk by excluding vulnerable crops on fields classified as highly erodible. Farmers are often well aware of the erosion risk on their fields, if only from experience. Most of them therefore know which fields to avoid when planting crops like sugar beet or potatoes.

Changing land use. Some farmers simply avoid the erosion risk by changing arable land with a high erosion risk into permanent grassland or by growing a perennial crop. The change is often done in response to a new regulation which may encourage the growing of fuel wood. Some of these regulations, however, can have negative effects. In some cases, set-aside regulations have led to farmers to use only fields with a high erosion risk, so that they can meet their obligation of leaving a certain percentage of their production land under fallow. In some cases farmers from outside the high erosion risk areas may even lease erodible land in place of their own set-aside.

Sowing deeper. Some measures are only focused on minimising the risk of crop damage without reducing the erosion risk. The main one of these is sowing deeper than normal. Germination will be slower, but the farmers who use this measure explain that the risk of losing the crop is reduced.

The adoption of soil conservation measures on arable land

The adoption of soil conservation measures to control wind erosion started in the USA during the 1930s (Fryrear this volume). In Europe, adoption by farmers depends mainly on the perception of the problem and the financial risk entailed. In the case of wind erosion this differs considerably and depends on:

- The value of the crops: damage to crops like sugar beet and potato leads to higher extra costs or a reduction of the crop production compared to damage to crops like barley or fodder maize.
- The perception of soil fertility: the loss of soil fertility is often not noticed, being masked by the use of fertilisers, and because of this very few farmers see it as a problem (WEELS Technical Report to the European Union Commission for the period: 01/01/1999 to 31/12/1999).
- Land ownership: Most farmers are not willing to undertake large investments on land they do not own. In Sweden, some of the degraded fields are leased and only used as 'set-aside'. No measures are taken to improve the soil and stop further degradation on these fields.
- The off-site effects and actors involved: the extent of dust damage and nuisance in the local community determines whether farmers feel obliged to avoid wind erosion.

Beside farmers' perception of the problem, local regulations and incentives also influence the use of different measures.

In southern Europe reference to wind erosion as a land degradation process is absent in most national action programs and there are no specific policies to prevent wind erosion on agricultural land. At a farm level, the perception of the wind erosion problem in semiarid farming areas, where there is a risk of wind erosion, is in general low (Gomes, 2000).

Existing European policy related to soil conservation

The main objectives of the Fifth Environmental Action Programme (*Towards Sustainability*) are to "... maintain the overall quality of life; to maintain continuing access to natural resources; to avoid lasting environmental damage; to consider as sustainable a development which meets the needs of the present without compromising the ability of future generations to meet their overall needs".

European legislation makes no specific reference to the control of wind erosion; there are no direct policy measures at a European level to control it; and few measures exist in individual Member States. It is arguable, nonetheless, that public funds should be used to reward farmers in return for measures taken to control wind erosion.

The existing policy measures for soil conservation in general are identified in this section and their relevance to wind erosion is explored.

Mandatory measures

Bundes-Bodenschutzgesetz in Gemany. Mandatory measures against soil erosion apply to farmers in Germany. They need to be taken in the context of the soil protection Act (Bundes-Bodenschutzgesetz 17-03-1998 [BGBI. I S. 502]). General objectives for good agricultural practice are developed at the national level by a committee on which scientists, extension officers, policy makers and farmers are represented. "Good Agricultural Practice" means that farmers are compelled to take precautionary measures to preserve soil fertility and the capacity of the soil. The agricultural extension service should instruct farmers about the principles of Good Agricultural Practice. In cases of off-site damage due to erosion, the farmer can be penalised if he can be shown not to have farmed according to the Good Agricultural Practice Code. On the other hand compensatory payments are offered where constraints need to be met that go beyond what is legally required.

Regionally directives in the Netherlands. Directives to combat wind erosion in the Netherlands are restricted to particular regions. In these regions it is forbidden to create a situation on arable land or desinfected land in which wind erosion can occur. Farmers are therefore obliged to take at least one measure from a prescribed list. In the so-called Veenkoloniën farmers only are obliged to take measures on desinfected land. These directives became valid in July 2001.

Exemption from mandatory measures apply to farmers who use animal manure in the Netherlands.

In general mandatory measures apply to farmers who use animal manure. They need to work manure under the ground soon after application. However, farmers in Veenkoloniën and Texel are exempted from such restrictions so that they can reduce damage from wind erosion. The areas subject to exemption are those that are the most vulnerable to wind erosion. In these regions, the application of animal manure as a surface protection layer is permitted in the spring (Figure 4), but only if the farmer depends on a rotation with 50% (starch) potatoes. The main reason for this exemption is that other products are too expensive for this low margin crop.

Figure 4. The prevention of wind erosion by the spraying of liquid cattle manure (at 15 t ha^{-1}) is still allowed in the Dutch reclaimed peat landscapes.

Voluntary measures

Most European countries do not have statutory controls, as in Germany or the Netherlands, and instead use voluntary measures to prevent wind erosion, promoted through extension services. Such a system applies in the UK, where soil conservation practices are covered in the "Codes of Good Agricultural Practice". The code is intended to guide farmers in their attempts to prevent harmful effects on their soils of farming practices (MAFF, 1998). The Soil Code contains a lists of measures, and of 'best practices'. Farmers, however, are not legally bound to manage their farm according to this Code.

Compensatory measures

In some countries compensatory payments are offered in the context of EU Regulations 2078/92 and 2080/92. Some of these payments, like payments for planting fuel wood on arable land, can contribute to the control of wind erosion.

Incentives for environmental purposes are a new policy instrument in the EU, and the single most important instrument is Regulation 2078/92, which became operational in 1993. The regulation aims to "encourage farmers to make undertakings regarding farming methods compatible with the requirements of environmental protection and maintenance of the countryside, and thereby to contribute to balancing the market; whereas the measures must compensate farmers for any income losses caused by reductions in output and/or increases in costs and for the part they play in improving the environment". In Central Aragon (NE Spain), for instance, some compensatory agri-environmental measures recently issued by the regional government (BOA 26/02/2001, Orden 20/02/2001) on the basis of the EU Regulation 2078/92 could contribute to wind erosion prevention on fallow land (Table 1).

Table 1: Compensatory agri-environmental measures that could indirectly contribute to wind erosion prevention in fallow lands of Central Aragon (NE Spain)

1 Agri-environmental measure	2 Farmer's commitments
- Improvement of traditional fallow: "environmental" fallow (basic)	- Stubble retention at least during the five months after harvest - Grazing limited to appropriate stocking rates
- Maintenance of stubble and straw chopping (complementary)	- Spreading of chopped straw and chaff at least over 50% of the field
- Extensification systems for flora and fauna protection (basic)	- Stubble burying delayed to a fixed date - Maintenance of boundary strips and areas of native vegetation at least in 1% of the field
- Legume seeding in fallow lands for bird feeding (complementary)	- Controlled grazing of legume crops
- Environment improvement and conservation (basic)	- Grazing limited to appropriate stocking rates

Farmers who adopt these measures are however obliged to comply with three of the Good Agricultural Practices for soil protection listed in documents issued at the national level: prohibition of straw burning, prohibition of traditional tilling in the slope direction and obligation to keep grazing within certain stocking rates based on local annual average precipitation (Annex I, BOE 13/01/2001, R.D. 4/2001, MAPA, Madrid).

In addition to the agri-environment measures under Regulation 2078/92, the EU also introduced measures for the afforestation of agricultural land in 1992, following the reform of the Common Agricultural Policy. Both measures are obligatory, which means that member states have to establish a national framework for their operation, which will be defined, but possibly differentiated for different regions.

Under Regulation 2080/92, which concerns forestry, member states may grant support for afforestation to farmers, to all other individuals and to forestry associations or co-operatives or other bodies which afforest agricultural land. The support may be granted to meet:

- costs of planting;
- costs of maintenance of the woodland over a period of (the first) five years;
- income losses in agriculture because of afforestation; and
- investments in woodland improvements, such as the provision of shelterbelts, fire-breaks, water points and forest roads, and the improvement of woodland under cork oaks.

These grants can help to reduce the wind erosion risk if they apply to:

- afforestation of arable land with high wind erosion risk
- improvement of the existing shelterbelt network

Geographisches Institut
der Universität Kiel

Discussion and conclusions

Measures to prevent wind erosion

Recent research, some of it presented at the European Conference on Wind Erosion on Agricultural Land held in Thetford, UK, in 2001, makes it clear that in regions prone to wind erosion in north-western Europe most farmers take some measures to prevent wind erosion. They make use of many of the measures described in this chapter, even some of the most expensive, like the planting of shelterbelts. The most successful implementation is in places where there has been co-operation between farmers, scientists, contract workers for technical assistance and governmental agents, as in Denmark. Most farmers, however, take relative simple measures, which may also serve other purposes. An example is furrow-pressing, which is now commonly practised on light sandy soils, although many farmers claim that they do this to improve the water availability, rather than to control erosion.

The effect of measures varies considerably. Some have only a short-term effect (Table 2) or protect arable crops only in the period when they are most vulnerable to wind erosion. Table 2 also shows the main restrictions of the different measures.

In the loamy-textured extensive cereal growing areas of southern Europe, wind erosion may occur slowly and its harmful effect on soil quality and productivity goes unnoticed for years, especially it is masked by mixing due to intensive soil tillage. In these areas sustainable soil conservation and management practices, such as conservation tillage and crop residue management, are the best strategies to prevent wind erosion. In central Aragon (NE Spain), for instance, reduced tillage has proved to be a practical form of soil management to control wind erosion during the fallow period (Gomes, 2000). At the same time this practice can considerably increase dustiness (Figure 5).

Figure 5. Dust rising during conventional tillage in central Aragon (NE Spain).

Table 2: Most common measures to minimise wind erosion risk in North-western Europe

Type of measures	Measures	Remarks
Measures that minimise the actual risk (short term effect)	- Catch crops and autumn-sown varieties	Needs to be sown before the end of October to develop a sufficient cover
	- Mixed cropping	After main crop is harvested, second crop remains on the field
	- Nurse or cover crop	More herbicides needed
	- Straw planting	Expensive
	- Organic protective layer (e.g. liquid manure; sewage sludge; sugar beet factory lime)	Use depends on availability, and regulations on the use of these products.
	- Synthetic stabilisers	Unsuitable on peat soils, or for low-value crops
	- Cultivation practice (e.g. minimum tillage; plough and press)	Not suitable for all crop or soil types
Measures that lower the potential risk (long term effect)	- Smaller fields	Increase in operational times and costs
	- Change of arable land to permanent pasture or woodland	Loss of agricultural production and farm income
	- Increasing the clay content to 8 – 10%	Suitable material must be available close-by
	- Wind barriers	High investment costs, and loss of productive land. Takes several years before they provide full protection.

The need for a uniform wind erosion policy in European member states

The large variation in the way that farmers across Europe deal with wind erosion can be explained by the lack of appropriate legislation. There is no specific legislation to combat wind erosion in most European countries and, except for the Netherlands and Germany, intervention by regional or national government is restricted to advice.

Although the area affected by wind erosion in Europe is small in area, and despite the varied socio-economic context of European countries, a uniform and effective wind erosion policy for the whole of the community is desirable, given the implications for sustainability and the off-farm problems wind erosion can create (as in the production of dust). This is possible if it is part of a more general policy on soil management. The most appropriate form is within a Code of Good Agricultural Practice (CGAP). The main objectives of this code could be:

- to preserve soil productivity for the future;
- to restore soil productivity of already degraded areas; and
- to avoid off-site damage

A basic set of objectives like these could be the basis for detailed regulations and measures adjusted to the regional context, and introduced perhaps by regional governments. But the EU as a whole is also obliged to have more regard for the consequences of their agricultural policy regarding these objectives. A new body may be needed to meet this requirement, which could also advise regional governments to make better use of existing regulations and in making new ones.

As a first step, educational programmes could be launched at national or regional scales, as they have already been in some countries. These highlight the threat of wind erosion to soil productivity and describe specific farming practices to address the problem.

References

Als, Chr., 1989. How to succeed in planting 900 KM of shelterbelts per year in a small country like Denmark? In: *Soil erosion protection measures in Europe*, Eds. Schwertmann, U., Rickson, R.J. and Auerswals, K., *Soil Technology Series* **1**, Catena Verlag, Cremlingen, 1-13.

Borstlap, L., 1999. *Farming with Wind Erosion; Case study on-site and off-site effects of wind erosion in Scania, Sweden*, thesis, International Land and Water management, Larenstein International Agricultural College, Velp, The Netherlands, 47 pp.

Eppink, L.A.A.J. and W.P. Spaan, 1989. Agricultural wind erosion control measures in the Netherlands. In: *Soil erosion protection measures in Europe*, Eds. Schwertmann, U., Rickson, R.J. and Auerswals, K., *Soil Technology Series* **1**, Catena Verlag, Cremlingen, 1-13.

Fry, G.L.A., 1994. The role of field margins in the landscape, in *Field Margins:intergrating agriculture and conservation*, Ed. Boatman, N., Proceedings of a symposium held at the University of Warwick, Coventry, 18-20 April, *SCI Monograph*, **58**, 31 - 40.

Gomes, L., Ed. 2000. *Wind erosion and loss of soil nutrients in semiarid Spain* (WELSONS). Final Report, European Commission Contract ENV4-CT95-0182, LISA, Paris, 62 pp.

Hudson, N., 1981. *Soil conservation.* Batsford, London, 324 pp.

MAFF, 1998. *Code of Good Agricultural Practice for the Protection of Soil.* Ministry of Agriculture, Fisheries and Food, Welsh Office Agriculture Department, 66pp.

Morgan, R.P.C.,1995. *Soil Erosion & Conservation.* Second Edition, Longman, 198 pp.

Nuberg, I.K., 1998. Effect of shelter on temperate crops: a review to define research for Australian conditions, *Agroforestry Systems* **41**, 3-34.

Prince, H.C. 1962. Pits and ponds in Norfolk, *Erdkunde*, **16** (1), 10-31.

Riksen, M.J.P.M. and de Graaff, J. 2001. On-site and off-site effects of wind erosion on European light soils. In: *Land Degradation and Development,* **12** (1), 1-11.

Schimmel, H., 1975. 'Atlantische woestijnen' de Veluwse zandverstuivingen, *Natuur en Landschap,* **29e** (1-2), 11-45.

WEELS Technical Report to the European Union Commission for the period: 01/01/1999 to 31/12/1999.

Zobeck, T.M. 1991. Soil properties affecting wind erosion, *Journal of Soil and Water Conservation*, **46** (2), 112-118.

5. Wind erosion: research and policy in the United States

Donald W. Fryrear, formerly United States Department of Agriculture, Big Spring, Texas.

Introduction

Wind erosion has plagued mankind since 4000 BC in Mesopotamia (Weiss, 1996), and 1150 BC in China (Dong *et al.*, 2000); it still does today (Fryrear, 1981). As early as 1790 in the eastern United States Deane used hedges and locust trees to reduce the effects of wind erosion on agricultural production, and by 1824 Drown was adding 50 to 80 mm of clay to protect a blowy sand soil surface (McDonald, 1971). They knew, as did people millennia ago and as many farmers know today, that wind erosion can damage plants, degrade soil productivity, reduce visibility, and lead to lower environmental quality.

Deane's and Drown's methods are still basic in wind-erosion control, but it is now known that crop residues, cover crops, and soil roughening are also effective (Fryrear, 1969). But adapting crop residue management and soil roughening practices to large-scale farming has required the development of special equipment. This followed in the tradition of Jethro Tull's horse-drawn cultivator and seed drill in 1731 in England, and James Small's 1784 invention of the iron plough (American Farm Bureau, 2001). These implements were essential if the Great Plains of the United States were to be farmed, but they are also essential to soil conservation. It may be true, as Agnew and Warren (1990) asserted, that the imprudent use of large farm machinery accelerated wind erosion in the Dust Bowl of the United States in the 1930s, but many farmers believe that large machinery used in a timely manner also allows the effective control wind erosion.

Farmers can be helped by government policies to do what they already know to be effective control methods. As Agnew and Warren (1990) maintained "ultimately most of the practices that lead to the degradation of arid lands can be traced not to the specific farmers but to poorly conceived policies".

The assessment of wind erosion

Wind erosion may have already been a problem in 1790 in the United States, but its widespread impact was not fully documented until the 1930s. A 1937 survey (USDA, 1937) reported:

> "Wind erosion in the United States has developed to the point where it has become a very serious national problem. It is serious not only from the standpoint of tremendous damage already done, but also from the standpoint of probable

future damage of much greater proportions, unless effective control measures are soon applied.

The conditions around innumerable farmsteads are pathetic. A common farm scene is one with high sand drifts filling yards, banked high against farm buildings, and partly or wholly covering farm machinery, woodpiles, tanks, troughs, shrubs and young trees. In the fields near by may be seen the stretches of hard, bare, unproductive subsoil and sand drifts piled along fencerows, across farm roads, and around Russian-thistles and other plants. Numerous livestock have died as the result of strangling, eating excessive amounts of grit, and from starvation, all associated directly or indirectly with wind erosion and drought.

Wind erosion is widespread over much of the cultivated lands of the Great Plains region extending from the Canadian border almost to the Gulf of Mexico."

From 1934, the United States Department of Agriculture (USDA) was required to make an annual estimate of land damaged by wind erosion in the Great Plains (Table 1) (Arnn, 1975).

Year	Hect.	Year	Hect.	Year	Hect.	Year	Hect.	Year	Hect.	Year	Hect.	Year	Hect.
1930		1940	1.5	1950	1.2	1960	0.9	1970	1.9	1980	5.0	1990	3.3
1		1	1.0	1	1.5	1	0.6	1	0.9	1	2.1	1	1.7
2		2	0.4	2	2.0	2	1.2	2	0.8	2	2.2	2	1.2
3		3	0.5	3	2.1	3	1.7	3	1.5	3	5.0	3	1.9
4	1.6	4	0.5	4	6.2	4	1.6	4	2.3	4	3.4	4	2.3
5	2.0	5	0.7	5	3.8	5	0.5	5	2.5	5	3.6		
6	2.0	6	0.9	6	4.2	6	1.0	6	3.2	6	2.4		
7	2.4	7	1.0	7	1.5	7	0.5	7	1.1	7	4.8		
8	3.2	8	1.6	8	1.3	8	0.4	8	1.2	8	5.8		
9	0.7	9	0.8	9	1.0	9	0.8	9	2.1	9	3.2		
Ave.	1.98		0.89		2.48		0.92		1.75		3.75		2.08

Table 1. Annual land damage in millions of hectares in the Great Plains between 1934 and 1995 from United States Department of Agriculture-NRCS reports.

In the 1950s more than 6.2 million hectares of land were damaged compared to 0.4 million in 1942 and again in 1968. Years with large land damage are normally years of below-normal rainfall. Government programs cannot influence winds or droughts, but they can influence farming methods. The land-damage information has been used to do just this (Dideriksen *et al.*, 1974).

Starting in 1982, the Soil Conservation Service (SCS) of USDA was requested to make an assessment of the natural resources in the United States. A Natural Resource Inventory (NRI) report is prepared and submitted to Congress every five years. For these inventories, models are used to estimate soil erosion. The inventories are used to determine the quantity of soil conservation funding to be allocated to individual states. In connection with these reports, the SCS prepared a map of the United States in 19 showing where wind erosion occurred on cropland (Figure 1). By far the most extensi region of wind erosion problems was the Great Plains, but every State had areas of win erodible soil. The 1997 map of areas with highly erodible cropland is similar to the 1988 map, but shows many additional areas in the Midwest (Figure 2).

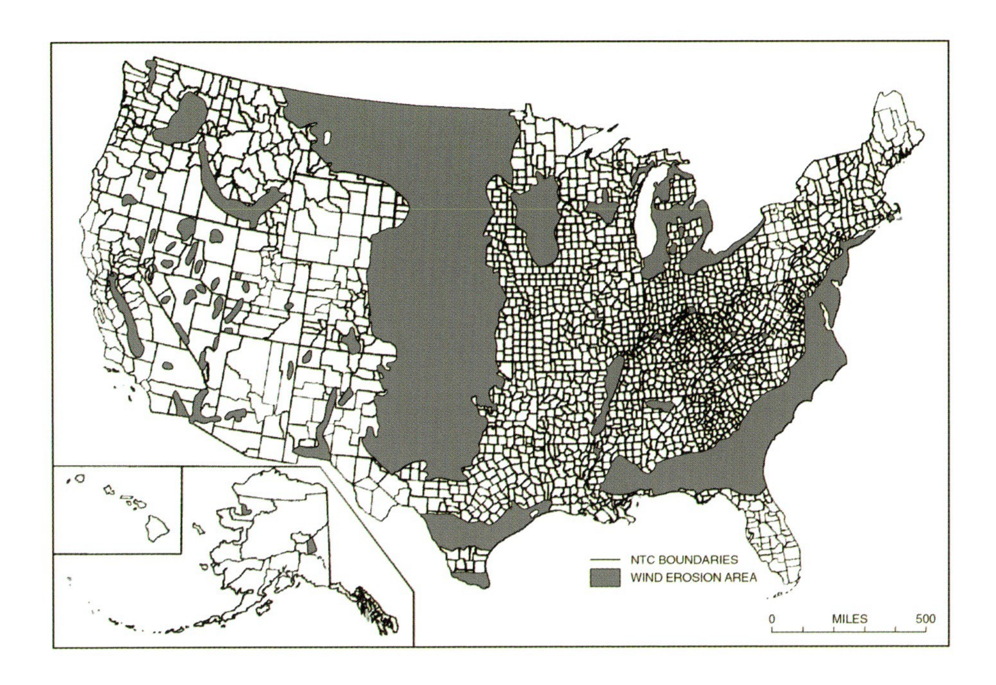

Figure 1. Areas where wind erosion occurs in the United States of America according to United States Department of Agriculture Soil Conservation Service map prepared in 1988.

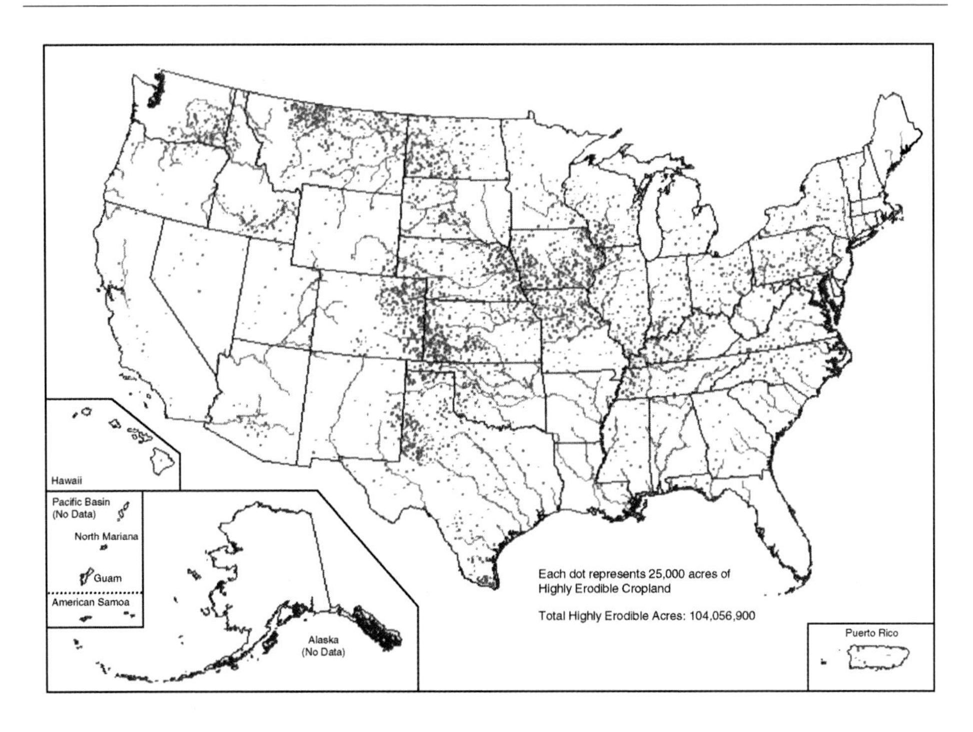

Figure 2. Highly erodible cropland in the United States of America from the United States Department of Agriculture-NRCS 1997 National Resources Inventory

Wind erosion research

Research continues to identify new techniques for controlling wind erosion and to expand our knowledge of the wind-erosion process (Fryrear *et al.*, 1998). Research can determine the reduction that is achievable without degrading the soil, crop, or atmospheric resources. Wind erosion control practices that reduce wind velocity or increase the threshold velocity of the soil include:

A. Covering or partially covering the soil surface with nonerodible material such as crop residues.
B. Increasing soil roughness with soil ridges or soil clods.
C. Reducing wind velocity at the soil surface by leaving crop residues standing in the field.
D. Reducing wind velocity at the soil surface with wind barriers.
E. Increasing threshold wind velocities for erosion with soil amendments.

The most effective practices combine these measures.

The applicability of these control practices and their combinations depends on local circumstance and weather conditions. A cover crop may not be an effective practice when drought might destroy the crop. Growing wind barriers in a region with highly variable erosive wind directions has limited benefit and may present management problems. Even the most effective control practices cannot completely halt erosion in exceptionally severe wind conditions. We cannot eliminate wind erosion, but when the most effective practices are used, we can reduce it.

Individual landowners probably funded the early research by Deane and Drown. The climate for research was completely changed in the first half of the 20th century, when wind erosion caused many farmers in the Great Plains to abandon their land and move to other areas (Chilcott 1931). The Federal Government then became the main research-funding agency, and by the 1990s most research was still funded by the federal government, mainly because wind erosion research involves a long term commitment, is high risk, but low return, and requires the careful training of research staff.

Policies

Efforts by government agencies to control wind erosion must be coordinated. Programmes must be complementary and not contradictory. The present list of the players in this process is given below and their relationships are shown in Figure 3.

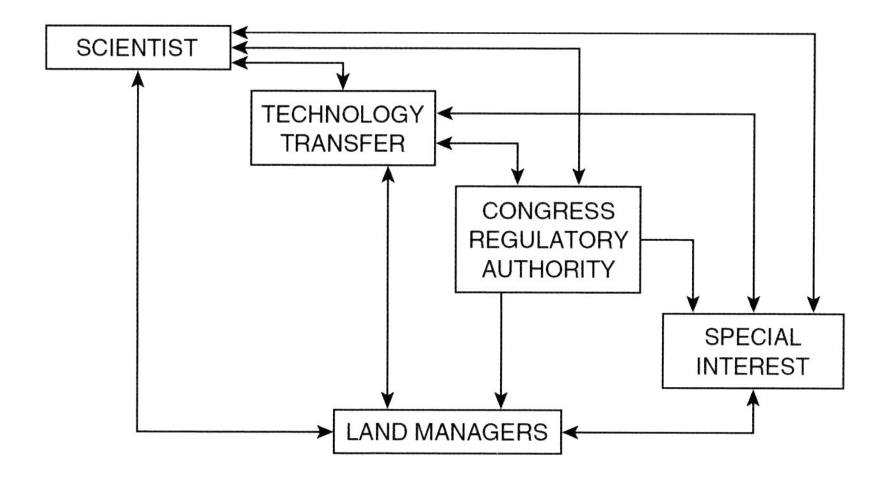

Figure 3. Players impacting wind erosion issues.

A. Scientists (anyone conducting meaningful research on the problem).
B. Land managers, including farmers, ranchers or agency personal responsible for production.
C. Congress/Regulatory/Policy makers including local, state, national or International bodies.
D. Special-interest groups include those interested in wildlife, environment, or finance.
E. Technology transfer groups includes extension and service agencies, or those giving farm services.

It is the land manager who puts the recommendations into practice. Thus his or her concerns, needs, and inputs are essential considerations if wind erosion control policies are to be successful.

Government policies concerning wind erosion have changed considerably over the years. Even in the 1930s, when wind erosion focused attention on the Great Plains, its problems were being attacked through Agricultural College Extension programs that encouraged diversification, better farming systems, crop and livestock improvement, cooperation, better markets (USDA, 1939). The following Federal Agencies were taking part:

AGRICULTURAL ADJUSTMENT ADMINISTRATION, designed to stabilize production and favourably influence prices,
SOIL CONSERVATION SERVICE designed to maintain the productivity of soils through the conservation of soil and water,
FARM CREDIT ADMINISTRATION, designed to provide credit cooperatively at reasonable rates of interest for financing farm operations,
FARM SECURITY ADMINISTRATION designed to stabilize low-income farm families on a self-sufficient basis,
FEDERAL CROP INSURANCE PROGRAM designed to compensate for crop failure by distributing the production of the farm uniformly over the year,
RURAL ELECTRIFICATION ADMINISTRATION designed to improve farm living conditions by the introduction of electric power at reasonable rates.

They were all needed. But, because there was essentially no wind erosion research being conducted, the scientists could not provide the information that the other groups needed. Wind erosion research was not funded at any government level or by private industry. Land managers were not trained or financially equipped to do wind erosion research. They were willing to try new techniques, but needed guidance on the most effective.

The demands of World War II, particularly the needs for food and fibre, and the return of favourable rainfall in the 1940s, overshadowed the wind erosion problem in the Great Plains. The agricultural sector responded, but rainfed agriculture was forever changed. Farm sizes doubled and tripled as more powerful farm tractors were developed and the sizes of implements increased. Crop rotations were replaced with single crop farming systems.

By the early 1950s overproduction was the primary concern. Congress passed legislation to place erodible lands in a Soil Bank to reduce overproduction, but also to decrease wind erosion and crop production. Fields in the Soil Bank remained in permanent grass for 10 to 15 years. There had been some research by this stage, and special interest groups were beginning to voice opinions about the role of agriculture in the production of dust, and the effects of this dust on the quality of life. The Soil Bank programme was reasonably successful for the life of the legislation, but it had not been designed to address long term problems associated with the farming of highly erodible soils in semiarid regions. Climatic conditions in the 1950s were as severe as those in the 1930s, but wind erosion in the Great Plains was considerably less because of improved technology, though erosion did increase (Table 1).

The 1985 Farm Act was directed at reducing excessive erosion, stabilizing land prices, and slowing the, by then, chronic overproduction. Under the Act wind erosion had to be estimated for all farmland. Fields with excessive erosion were retired or required to be treated with acceptable conservation practices. Many farmers retired their highly erodible lands for a period of 10 years. They "bid-in" their fields and if the field met the erosion requirements, and funds were available, bids would be accepted. Not all fields that were entered were accepted, but 38 million acres were. The requirement for acceptance was reduced erosion, and for this purpose the Wind Erosion Equation (WEQ) was used for estimating wind erosion (Woodruff & Siddoway, 1965). Land managers were also encouraged to use soil-conserving practices such as surface mulches, wind barriers, and soil roughening if they were growing crops supported by government programmes. The technology transfer group provided information on the management practices that would be acceptable.

Unfortunately, new information developed between 1965 and 1985 was not applied until much later. Moreover, wind erosion research initiated after the 1985 Farm Bill revealed that WEQ erosion estimates were considerably different from field measurements. Major differences were noted for unusually large or small fields and for rainfall regions significantly different from that of western Kansas, where the WEQ had been developed. The WEQ had been designed to illustrate the influence of various conservation practices on wind erosion. In order to use it as a regulatory tool a simple method of determining maximum allowable erosion was developed. Fields that had an "*EI*" greater than eight were considered highly erodible. "*EI*" is the Climatic Factor (*C*), times soil erodibility (*I*), divided by tolerable soil loss (*T*). A critical evaluation of *C, I*, and *T* raised concerns about the value of *EI*. Arbitrarily to establish an *EI* of eight for all soils had little scientific merit.

The land damage reported by Soil Conservation Service in the 1980s was considerably greater than for other decades (Table 1). While this may be a reflection of additional climatic hazards, the magnitude of the increase would suggest that the recommended conservation practices were not effective. The practices recommended by the 1985 Farm Bill were not implemented until about 1988, and appeared then to have made a difference, because there was a decline in land damage in the 1990s.

In the 1996 Farm Act, Congress sought a greater emphasis on environmental benefits, largely because wildlife groups sought larger blocks of land for wildlife habitat. Under this Act, the bids from land managers to put land into conservation reserve were compared with the Environmental Benefits Index (*EBI*), in which points were given for wildlife habitat benefits (100), water quality benefits (100), on-farm benefits from reduced erosion (100), long-term benefits such as from planting trees (50), air quality benefits from reduced erosion (35), and benefits of enrolment in conservation priority areas to help to improve adverse water quality, wildlife habitat, or air quality (25). Additional points were given for providing wildlife cover, for benefiting endangered species, being near protected areas, or for restoring wetlands that provided nesting habitat. In the 15th sign-up in March 1997, all bids that exceeded an *EBI* of 259 were accepted. This legislation emphasized the inseparability of agriculture from the rest of society. It acknowledged that what the land manager did to his land impacted the environment and wildlife. The most efficient and effective control practices occurred when all parties communicated their interest and concerns.

There are problems, nonetheless. Under the present programs communication between land managers and special interest groups is not perfect and inputs from land managers to legislation are few. To maximize the benefits of new technology and to insure that legislative and research efforts are effective and cost-efficient, technology transfer must serve as an independent unbiased clearinghouse of information. All the interested players should communicate before legislation is prepared and provide feedback on strengths and weakness after the new legislation has been enacted.

Objectives of research and government policies

The first objective of wind erosion research is to understand the wind erosion process. The second objective is to use this knowledge to design effective and efficient control practices. The third objective is to identify the limits of various control practices. Research programmes should relay new knowledge to the technology transfer group and should be prepared to answer inquires from the other players. In the process of answering inquires, new areas of funding and support may surface.

If there is effective exchange between scientists, land managers, the technology transfer group, the legislature and special interest groups, defendable practices can be included in government programs. The 1996 Farm Act was a compromise between the demands of environmental groups who sought greater emphasis on environmental benefits, wildlife groups who sought wildlife habitat, and land managers who wanted minimal modification of the existing legislation. But if legislation is made in response to special interests without recognizing its impact on the land managers, effective control of wind erosion may not occur. Legislation would be better if lines of communication between the various players were improved.

Conclusions

Tremendous advances have been made in understanding the wind erosion process and in developing effective wind erosion control practices in the United States. The United States Department of Agriculture has funded most of the research. In recent years, because of the interest in erosion, airborne dust, and improved resource utilization, support for wind erosion research has come from those interested in environmental sustainability and wildlife.

Nonetheless, the data show a continuing problem. To minimize the damage will require the combined efforts of conservation-minded land managers and more effective legislation. No single system will be suitable for the entire country because of the diverse cropping and management systems.

Wind erosion can never be eliminated, but effective control practices can minimize the problem. Continued improvement in control strategies need long-term commitments of research funding. As public awareness of air quality becomes keener, the interest base will expand and additional sources of funding may surface. New wind erosion control practices will evolve that are compatible with modern farming systems, and are in harmony with the environment.

References

Agnew, C.T. and Warren, A. 1990. Sand trap: agriculture not desert is the greatest threat to arid land, *The Sciences* (New York), March/ April, 14-19. also: 1993. The sand trap, in *The culture of science*, Eds. Hatton, J. and Plouffe, P.B., Macmillan, New York, 517-525.

American Farm Bureau 2001. Significant events in agricultural industry history, *Resources*, ASAB, **8** (1), 18.

Arnn, John W. 1975. *The New Wind Erosion Damage Reporting System*, Paper presented to Soil Conservation Service-Texas Tech University Conservation Workshop, July 14-16, 1975.

Chilcott, E. C. 1931. *Crop rotation and tillage methods, United States Department of Agriculture Miscellaneous Circular*, **81**, *Supplement*, 1, 164 pp.

Dideriksen, R.I., Robinson, A.R., Houseman, E.E., Hill, H.L. and Fedkiw, J. 1974. *Appraisal of SCS Wind Erosion Damage Assessment and Reporting and alternatives for improved damage assessment*, United States Department of Agriculture, September.

Dong, Z.B., Wang, X.M. and Liu, L.Y. 2000. Wind erosion in arid and semiarid China: an overview, *Journal of Soil and Water Conservation*, **55** (4), 439-444.

Fryrear, D.W. 1969. *Reducing wind erosion on the southern Great Plains*, **MP-929**, Texas A&M University.

Fryrear, D.W. 1981b. Dust storms in the southern Great Plains, *Transactions of the American Society of Agricultural Engineers*, **24** (4), 991-994.

Fryrear, D.W., Saleh, A., Bilbro, J.D., Schomberg, H.M., Stout, J.E. and Zobek, T.M. 1998. *Revised wind erosion equation, Technical Bulletin*, **1**, Wind and Water Erosion Conservation Research Unit, United States Department of Agriculture-ARS, Southern Plains Area Cropping Systems Research Laboratory.

Hedin, L.O. and Likens, G.E. 1996. Atmospheric dust and acid rain, *Scientific American*, **275** (December, 6), 56-61.

McDonald, A. 1941. *Early American soil conservationists, Miscellaneous Publication*, **449**, United States Department of Agriculture, Soil Conservation Service, 62 pp.

Piper, S. and Lee, L.K.. 1989. *Estimating the offsite household damages from wind erosion in the western United States, ERS Staff Report*, **89-26**, Economic Research Service, Resources and Technology Division, United States Department of Agriculture, Washington DC, 21 pp.

United States Department of Agriculture. 1937. *Soil conservation reconnaissance survey of the southern Great Plains Wind-Erosion Area, Technical Bulletin*, **556**.

United States Department of Agriculture. 1939. *Conference tour, southern Great Plains*, H.V. Geib, Soil Conservation Service, Field Representative. August 26 to September 3, 1939.

Weiss, H. 1996. Desert storm, *The Sciences*, **36** (3), 30-36.

Woodruff, N.P. and Siddoway, F.H. 1965. A wind erosion equation, *Proceedings of the Soil Science Society of America*, **29** (5), 602-608.

6. How researchers, policy makers and the community work together to minimise wind erosion in Australia

John Leys, Centre for Natural Resources, Department of Land and Water Conservation, Gunnedah, NSW, Australia

Introduction

Research, policy, funding and the implementation of land management practices are now strongly linked in Australia. Problems are prioritised and funded by both government and industry. The government's agenda is generally more environmentally-orientated while industry's is more production-orientated. This can be adversarial, but with a decline in government expenditure, there has been a redirection of government funding towards projects in which both industry and the community and involved.

Government now actively encourages communities to participate in the planning and implementation of better in natural resource management. The emphasis is on creating partnerships between government, communities and private industry. This Australian model of natural resource management has few parallels globally (Polkinghorne 1999). The programme that is best known for its community involvement is Landcare, although there are several other similar programmes in Australia. Government now funds strategic research that will either directly support government policy, as when it may provide evidence for a decision, or it forms partnerships with industry that encourage it to change its land management practices.

It is true that funding bodies generally direct the nature and focus of the research. This is especially the case with applied research. Examples of programmes funded by industry include those of the Grains Research and Development Corporation (www.grdc.com.au) and of those by the federal government include of the Natural Heritage Trust (www.nht.gov.au). These programmes have been established to implement government or industrial policy. More basic research is funded by the Australian Research Council (www.arc.gov.au), which has a general call for funding after which each project is assessed more on its scientific merit than on its contribution to practical problems.

Thus the current research model is a funding partnership between government, community and private industry. Its success lies in: the commitment of each group to an agreed goal; the skills and knowledge from within the group (or their ability to attract them) and the resources and funding that each group can contribute. Each project requires cooperation and empowerment. This is a move away from the model in which experts (generally in government) did the research and the planning, the government undertook the transfer of the policy to the community through an extension programme, and the community (the farmers) implemented the policy. The scientists were then far removed from the land managers.

The new model means that the scientists, policy makers, government extension staff, farmers and the funding body all work together throughout the entire project. This paper discusses this process. The perspective is from the State of New South Wales. The first part of the paper outlines the place of wind erosion in Australian environmental issues and the policy that exists to mitigate its effects. The next section outlines how science, policy and community were integrated in the LandCare program. The third section details two projects that demonstrate the process. Finally the paper concludes with some comments on why the current approach is successful and some of its problems.

Wind Erosion as an Environmental Issue

Wind erosion was certainly an issue behind the first soil conservation act in Australia, which was passed by the New South Wales Government in 1938 (Northrop 1999), largely because it had provided powerful images of the "unhealthyness" of the environment. When a dust storm later rolled over a city like Melbourne in February 1983 (Raupach *et al.* 1994), wind erosion again became an issue for the community and the government. And when the incidence of hospital admissions due to respiratory disease increased dramatically, as it did in Adelaide in May 1994, the public again became anxious about wind erosion. But incidents like these are too sporadic for it to have a high profile. Land and water salinisation, loss of biodiversity, and water quality are much more prominent in the natural resource agenda at present.

Nonetheless, there has been a small, but consistent thread of funding for wind erosion research in Australia for the last 30 years. It has significantly expanded our understanding of the processes and impacts involved (Houghton and Charman 1986, Leys 1998, McTainsh *et al.* 1989) and brought about a real fall in the incidence of wind erosion over most of Australia (McTainsh 1998)(Figure 1). One major reason for this decline is the improvement in land management that can be linked to research programs that demonstrated the impacts of erosion on the soil (Leys and McTainsh 1994). Other reasons include:

- Changes in government policy and legislation directly aimed at such change; and
- Increased awareness among land managers of the link between agriculture and environmental sustainability.

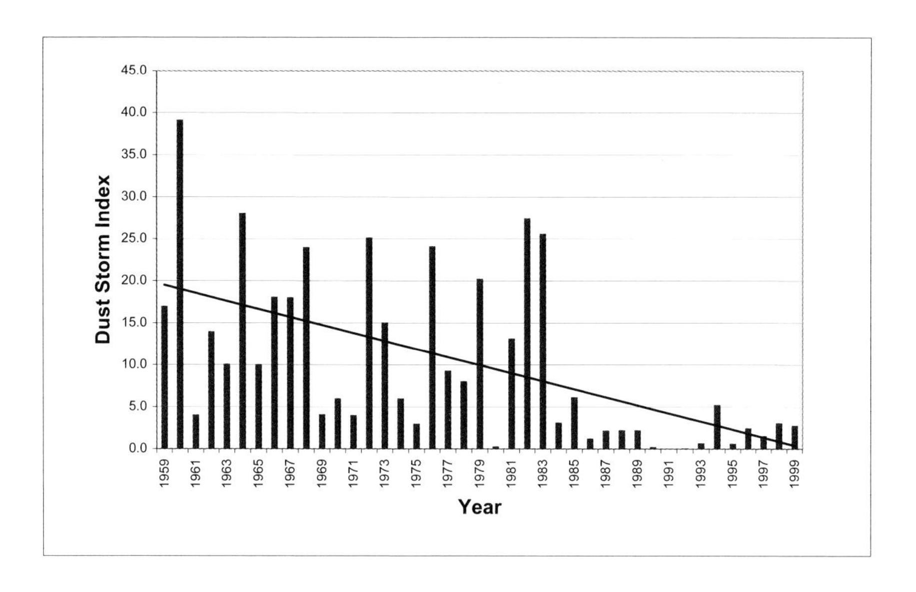

Figure 1: Dust storm index (DSI) at Mildura in South Eastern Australia for the period 1959-1999. Trend line shows a decline in DSI over the last 40 years (source McTainsh pers comm)

Wind Erosion Policy

The Commonwealth Government has never legislated with respect to soil conservation despite having the power to do so in the constitution (Australia 1986). It is in the states where this kind of legislation has been enacted (as in the Soil Conservation Act of NSW, 1938; the West Australian Soil and Land Conservation Act, 1982; and the South Australian Soil Conservation and Landcare Act, 1989), but this legislation has had little effect, largely because it had few specific targets or objectives and inadequate provisions (Hannam 2000). The policy that has been developed for soil conservation has been closely tied to agricultural production, but it has been confused about whether soil conservation is a "conservation" or "agricultural" activity (Hannam 2000).

An example of a state-based soils policy is New South Wales' "State Soils Policy" (Soil Conservation Service of New South Wales 1987). It was a major initiative, but was weakened by not being linked to the specific objectives or to the provisions of the earlier New South Wales Soil Conservation Act, 1938 (Hannam 2000). These weaknesses and a lack of targets for changing attitudes to soil conservation meant that the policy was not widely implemented. The policy (http://www.dlwc.nsw.gov.au/care/soil/policy.htm) is therefore little more that a general government statement that "soil should be looked after" rather than anything that sets out strong policy guidelines.

While there is legislation and some policy on soil conservation in general, there is no focused governmental "policy" on wind erosion *per se*. There is, nonetheless, considerable government and non-government activity that contributes towards the control of wind erosion. The legislative framework and government programmes that impact on wind erosion fall into three major areas: soil management; vegetation management; and support for community planning and participation.

There is more policy and legislation for the management of vegetation within each state. The Native Vegetation Conservation Act and the Draft Vegetation Conservation Strategy of NSW are examples (http://www.dlwc.nsw.gov.au/care/veg/conserv_strategy/index.html). In both of these, biodiversity is given more attention than soil conservation, and when they do relate to soil conservation, salinity is given more priority than erosion. However the general approach for both is to conserve vegetation for all its values, one of which is soil conservation.

Thus, the most significant contribution that government in Australia has made towards the protection of the soil resource is in the area of support and financial assistance for community participation and planning, as in Integrated Catchment Management:
(http://www.dlwc.nsw.gov.au/care/cmb/index.html) and Landcare
(http://www.dlwc.nsw.gov.au/community/landcare/index.html).

Government now uses funding strategically to implement policy. At the federal level, the Departments of Environment Australia and Agriculture Fisheries and Forestry Australia administer the Natural Heritage Trust (NHT)(see www.nht.gov.au). The Trust aims to integrate biodiversity conservation and sustainable agriculture and to provide funds for effective, on-ground activity where it will really deliver, empowering Australians to overcome local and regional environmental problems. Within the Department of Agriculture Fisheries and Forestry Australia, lies the Bureau of Rural Sciences, which covers both science and policy and which aims to develop integrated evidence-based policy development (see www.affa.gov.au/ruralscience.html). NHT has many programmes (http://www.nht.gov.au/programs/index.html), of which the one of most interest here is the Landcare program.

Landcare – linking government, communities and private industry.

The catalyst for a change in approach from the traditional management of natural resources by government came about in the 1980s when 50 years after the introduction of the Soil Conservation Act (1938) in the state of New South Wales, there was still seen to be significant soil degradation (Soil Conservation Service, 1989; http://www.dlwc.nsw.gov.au/care/soil/landdeg/contents.pdf). Although good progress had been made after the Soil Conservation Act was passed, it was apparent that by relying only on government initiatives, the rate of progress would not be adequate to guarantee conservation (Northrop 1999). What was required was more action than government alone could provide. What was required was community ownership of soil conservation and this came about in 1984.

The process of moving the responsibility of managing natural resources from the government to communities began in 1983. Although the responsibility for soil conservation in Australia lay with the State governments, the Federal Minister for Primary Industry wanted to see a more national approach. Federal money ($A1 million) was set aside for the National Soil Conservation Program (NSCP), which sought to develop and implement national polices for rehabilitation and sustainable use of the nation's soil and land resources (Northrop 1999). At first, most of the applications came from state government agencies. Community involvement in projects was first sought in 1984 when two landholders in the state of Victoria sought funding to address soil erosion issues in their catchments. In essence they wanted control of the funding to do what they thought was best. The numbers of this type of project quickly increased as community groups saw a way of funding projects that were of interest to them and not necessarily to state government departments. As the applications for this kind of project increased, the focus moved from water erosion, to other forms of land degradation (salinity, soil structure decline, wind erosion and so on). By 1986, the guidelines covered all forms of land degradation.

In 1985, the minister for Natural Resources in Victoria directed her department to involve the community with natural resource experts in the planning and implementation phases. This replaced a system in which the experts planned and the farmers performed the physical work. By 1986, LandCare Victoria had been formed as a partnership between the Victorian Farmers Federation and the state government. Significantly, NSCP had also changed its emphasis from dealing with the biophysical dimensions of land degradation to include economic and social issues. By 1987, the program had expanded to cover more than soil conservation and the term LandCare was adopted nationally with the agreement of the Victorians (who had originally registered the name).

By 1990, NSCP had funded 416 projects of which 111 had been undertaken by community groups. Also in 1990, the federal Prime Minister announced that the federal government would commit $A 26 million per year to the programme for the next ten years. In 1997, $A1.5 billion was invested in the Natural Heritage Trust by the federal government to ensure on-going funding for community-based programs (http://www.nht.gov.au/overview.html). By the year 2000, Landcare had 80% national awareness and had generated $A 58 million in sponsorship and media coverage (Scarsbrick 1999). There were over 4,000 LandCare groups and had agreement from all the political parties, conservation and farmer groups, urban and rural communities. The LandCare movement had also shown the importance of whole-farm and catchment planning and implementation that was based on networks of local groups, and not imposed sub-regional, state or national level authorities (Kirner, 2000; www.nre.vic.gov.au/conf/landcare2000/pdf's/Kirner Joan.pdf).

Examples of collaborative funding

Mallee Sustainable Farming Project Inc. In September 1998 in Werrimull, Victoria (population 35) in the dry mallee country (< 350 mm mean annual rainfall) of southeastern Australia a meeting of dryland grain production stakeholders was organised by Neil Smith, who was a member of the Grains Research and Development Corporation (GRDC; www.grdc.com.au). He asked the farmers, extension and research staff: "why were conservation farming methods not widely adopted in the dry mallee country?"

He explained that the mallee was still synonymous with dust storms, and that this was not an environmental image with which the grains industry wished to be associated. He then proposed that a farmer-led research and extension project could be funded, but only if the local farmers themselves wanted to take on the job of increasing the rate of adoption of conservation farming systems in the dry mallee. He offered $A 2 million (half from the GRDC and half from the federal government) to start the project and indicated that the state natural resource management agencies and the Commonwealth Scientific Industrial Research Organisation (CSIRO) were willing to participate. The result of this meeting was the formation of the Mallee Sustainable Farming Project Inc. The project is run by the farmers and is collaborative, and now has had over $A5 million in funding. The goal of the project is to "increase the adoption of sustainable and profitable farming practices in the Mallee of

South Australia, New South Wales and Victoria" (www.msfp.org.au). The project now reaches 1600 farmers over an area of about 100,000 km^2.

The project is now highly collaborative, involving funding and resources from federal (CSIRO, NHT Landcare program, University of South Australia, Soil and Land Management Cooperative Research Centre), the state government (Department of Land and Water Conservation, NSW Agriculture, Natural Resources and Environment, Primary Industries and Resources South Australia), the agricultural industry (GRDC) and private business (Pivot Agriculture, RDTS Cereals). One thing that distinguishes the project from other GRDC farming systems projects is its inclusion of environmental issues (wind erosion and ground water recharge). The links between research, policy, funding and the implementation of land management practices that minimise wind erosion are all strongly demonstrated in the project. All services to the project (research, extension, publicity etc) are commissioned by a farmer's committee, which has the explicit goal of increasing the adoption of sustainable and profitable farming practices.

The project has several sub-projects:

- Core trial sites: where intensive scientific experiments are conducted to identify the processes that are operating to limit crop production and the level of impact of farming systems on wind erosion and soil water use.
- Focus paddocks: 45 paddocks are monitored for agronomic and environmental performance (wind erosion and groundwater recharge)
- Benchmarking: sites at which surveys are conducted by farmers every year to assess attitudes and changes in farming practices
- Machinery performance: sites at which a range of tillage machinery is evaluated for soil disturbance, seed placement and yield
- Soil biology: where the micro-biological elements of the farming systems are investigated to identify limitations to nutrient cycling and crop growth

This project has two more years to run on its current funding, but looks like being funded for several years after that. It is an excellent example of community-based research, using industry and government funding to address industrial and governmental policies by influencing changes in land management practices.

Minimising the impact of pesticides in the riverine environment. This programme arose out of concerns about the health of rivers and about the potential for adverse economic impact of the regulation of the cotton industry in eastern Australia that might be based on unsubstantiated evidence. Anecdotal evidence had linked fish kills to the pesticides used in cotton growing, but the community felt there should be no rush into a strong regulatory mode without better evidence. A collaborative program was established between the major stakeholders with a $A6 million budget of which half came from the research and development corporations. The stakeholders included two federal funding agencies with interests in natural resource management (the Land and Water Resources Research and Development Corporation, LWRRDC, www.lwrrdc.gov.au; and the Murray-Darling Basin Commission, MDBC; www.mdbc.gov.au), industrial partners (the Cotton Research and Development Corporation, CRDC) and irrigators, plus the relevant state agencies, such as the NSW Departments of Agriculture, Land and Water Conservation, Environmental Protection Agency.

After a review of existing knowledge, and the identification of priorities by the stakeholders, a research and development programme was established. It commissioned research on the fundamental processes of the movement of pesticides and their impact on the riparian environment. One process that was to be researched was the movement of endosulfan on dust from farm to river. After the research was completed (Larney *et al.* 1999, Leys *et al.* 1998) potential solutions were identified for minimising this movement and these were incorporated into a "best-practice" manual for the cotton industry. The program was completed by launching a pesticide best-practice initiative (Williams 1997) and the results and model of collaboration were disseminated to other industries and the wider community (Schofield 1999).

This program highlights the link between research and management in a strategic policy, funded by all stakeholders. Wind erosion research may have been only a small part of the whole programme, but this account shows the nature of the current funding environment for research in Australia in which there are strong links between research, policy and the implementation of practices that minimise wind erosion.

How community participation brings policy and science together

It makes sense to integrate activities that seek similar outcomes. Strategically applied funding can produce a more integrated approach to resource management. Communities like to be involved in the process making and implementing land management decisions and the research and policy people are encouraged into the partnership by funding opportunities. The key issues that have encouraged the new approach are:

- Land managers prefer consultation and involvement to regulation
- Governments, for cost reasons, prefer to utilise the social and human capital available in the community rather than to supply all the resources

- Both government and community agree that the final outcomes tend to be more lasting than those of the older approaches because communities have greater ownership of the on-ground works
- Because LandCare and the NHT are national programmes which cover urban, coastal and rural areas through a range of projects from biodiversity conservation to sustainable agriculture, many communities can participate
- With the full community involvement, support from urban areas flows to rural areas
- The LandCare model adds a new social dimension to the process, which discourages rural depopulation.

Despite the success of community involvement in natural resource management, there have been and remain many hurdles.

Problems with community participation

Funding. One of the biggest impediments to community participation is lack of long term funding. Funds that continue for periods of the order of decades are required because land degradation occurs slowly and may have impacts many years after it first occurs. Only the assurance of long-term funding will empower communities to undertake projects that matter to them. However, some initial funding can attract other partners (government agencies, industry, agri-business), as the case studies have shown. It is because of this that the federal government is committed to long term funding because of the magnitude of environmental problems and their impacts on the whole economy (www.nre.vic.gov.au/conf/landcare2000/pdf's/Govt Response 2 Review.pdf).

Time. Community-based committees take more time to plan and implement a project than in the former top-down approach because few are skilled in project management. This can frustrate government and industrial partners, but is essential to success. Coordinators can help in: prioritisation; the development of strategic plans; and in the review the project as it progresses.

Environmental focus. Many younger farmers are more focused on production than on environmental issues. The Mallee Sustainable Farming Project avoids this problem by clearly stating the link between production and environmental sustainability.

Accountability. LandCare groups find it difficult to demonstrate the cost benefits of their projects at catchment and regional scales. There is a belief in some quarters that investment in community groups is not justified because the money invested does not return adequate environmental gains. Despite this, the government acknowledges that a great deal has been achieved in:

- Raising the level of investment in the natural environment,
- Adding value to the contribution of other community and State government stakeholders, and
- Raising community awareness and empowering communities to create new social networks to facilitate cooperative activity across regions (www.nre.vic.gov.au/conf/landcare2000/ pdf's/Govt Response 2 Review.pdf)

It is because of these arguments that government will continue to support community-based projects.

Ongoing Challenges. As an example of a community program that has been operating for over a decade now, we see that Landcare faces some major challenges:

- Complacency. LandCare just did not happen; it took an immense effort. Complac would put the programme at risk
- The need for continuing support of the volunteers and paid staff
- The temptation of change for change's sake; the perception that what is new is always better
- Underestimation of the role of local groups; the participation of local groups, in the decision-making as well as the work, though time consuming, is essential. (www.nre.vic.gov.au/conf/landcare2000/pdf's/Kirner Joan.pdf).

Conclusions

The integration of research, policy and implementation in land management to minimise wind erosion in Australia has been achieved by strategic funding by government and industry. The desire by communities to be responsible for land management has also been vital. With commitments from all partners, significant advancements have been made in controlling wind erosion.

Note

The views expressed by the author do not directly represent those of the Department of Land and Water Conservation, or of its sponsors.

Acknowledgements

I would like to acknowledge: The European Unions COST 623 programme which enabled me to speak at the COST 623 meeting "European Conference on Wind Erosion on Agricultural Land"; Andrew Warren and Geert Sterk for inviting me to do this; Grant McTainsh for the dust storm index data; Warren Tierney and Ian Hannam for their comments on the extent of wind erosion policy and the Board of Directors of the Mallee Sustainable Farming Project for allowing me to participate in an truly exceptional program.

References

Australia 1986. *The Constitution as altered to 31 October, 1986*, Australian Government Printing Service, Canberra.

Hannam, I.D. 2000. Soil conservation policies in Australia: successes, failures and requirements for ecologically sustainable policy, in Soil and water conservation policies and programs, Eds. Napier, T.L., Napier, S.M. and Tvrdon, J., Czech Agricultural University,

Prague / CRC, Boca Raton, 493-514.

Houghton, P.D. and Charman, P.E.V. 1986. *Glossary of terms used in soil conservation*, Soil Conservation Service, Sydney.

Larney, F.J., Leys, J.F., Müller, J. and McTainsh, G.H. 1999. Dust and endosulfan deposition in a cotton-growing area of northern New South Wales, Australia, *Journal of Environmental Quality*, **28** (2), 692-701.

Leys, J.F. 1998. *Wind erosion processes and sediments in southeastern Australia*, PhD thesis. Griffith University, Brisbane, 277 pp.

Leys, J.F., Larney, F.J., Müller, J.F., Raupach, M.R., McTainsh, G.H. and Lynch, A.W. 1998. Anthropogenic dust and endosulfan emissions on a cotton farm in northern New South Wales, Australia, *The Science of the Total Environment*, **220**, 55-70.

Leys, J.F. and McTainsh, G.H. 1994. Soil loss and nutrient decline by wind erosion - Cause for concern. *Australian Journal of Soil and Water Conservation*, **7** (3), 30-40.

McTainsh, G.H. 1998. Dust storm index, in *Sustainable agriculture: assessing Australia's recent performance*, *SCARM Report*, **70**, Standing Committee on Agriculture and Resource Management, Canberra. 55-62.

McTainsh, G.H., Burgess, R. and Pitblado, J.R. 1989. Aridity, drought and dust storms in Australia (1960-84), *Journal of Arid Environments*, **16** (1), 11-22.

Northrop, L. 1999. Looking back on the origins of the national LandCare program, *Australian LandCare*, December, 17-18.

Polkinghorne, L. 1999. Decade of LandCare success, *Australian LandCare*, December, 19.

Raupach, M.R., McTainsh, G.H. and Leys, J.F. 1994. Estimates of dust mass in recent major dust storms, *Australian Journal of Soil and Water Conservation*, **7** (3), 20-24.

Scarsbrick, B. 1999. Ten years of progress ushers in our century, *Australian LandCare*, December, 21.

Schofield, N. 1999. Origins and design of the cotton pesticides program, in *Minimising the impact of pesticides on the riverine environment: key findings from research with the cotton industry*, Eds. Schofield, N. and Edge, V., *LWRRDC Occasional Paper*, **23/98**, Canberra, Australia, (Land and Water Resources Research and Development Corporation, Canberra, Australia).

Soil Conservation Service of New South Wales 1987. *State soils policy*, SCSNSW, Sydney.

Williams, A. 1997. *Australian cotton industry best management practices manual*, Cotton Research and Development Corporation, Narrabri, NSW.

7. The future of wind erosion research and policy in Europe

Reported by Andrew Warren, University College, London

Introduction

This section summarises the outcome of discussions at the final meeting of ECOWEAL (European Conference on Wind Erosion on Agricultural Land) at Thetford in May 2001. The meeting was the first to discuss wind erosion in Europe. It was attended by policy-makers (from national agricultural departments and the EU itself) and scientists. As such, it was a unique opportunity to discuss a framework for research and policy. The framework should be seen as a discussion document that might inform:

- The coming deliberations in Europe on soil policy in general.
- The development of policy for research on wind erosion in the EU in more general terms
- The coordination of policy between national and European organisations

We chose to discuss only issues that were of European-level concern. These are the issues that we believed might benefit from coordinated research. They include issues that:

1. Gain from cooperation between the greater mass of scientists at the European than at the national level. Wind erosion is generally a minor concern in most national frameworks, so that there are few wind erosion scientists or policy-makers familiar with wind erosion in each country. These groups only gain credibility when amassed at the European scale.
2. Gain from coordinated investment and division of labour among specialists in Europe.
3. Are transboundary issues, such as the transport of dust that must be considered at the supra-national level.

We first discussed issues in two categories:

- ***Those already well developed in Europe***. These are emphatically not issues in which there should be disinvestment, for techniques in all the areas are far from being perfectly covered. They are issues that are at present well funded, or which have reached some plateau of achievement.

- ***Those issues that should now be research priorities.*** These include scientific and policy research

What we already do well

- Understanding processes, including the quantification and monitoring of sand and dust transport.
- Risk assessment at local and regional scales
- Control measures
- Creating awareness of wind erosion
- Interdisciplinary research

Research priorities

- Research on the production and diffusion of dust generated by agricultural practices

  This includes, the measurement (instrumentation and sampling), and modelling of dust emissions and diffusion from agricultural fields in relation to weather, climate, cropping pattern, tillage practice, etc.

- The development and application of wind-erosion models

  Wile there are now excellent tools for running and displaying models, especially in space (using GIS technology), these need to be tested and compared using comparable data sets, and used to test different land use, policy and climate-change scenarios.

- Standardization of monitoring and data handling

  The EU is in a good position to ensure that data concerning wind erosion and dust measurement from across Europe are collected with comparable techniques (instruments, spatial and temporal sampling frameworks, data handling and storage formats etc.)

- Control method evaluation and demonstration with farmers

  While many control measures have been shown to be potentially applicable (see above), there still needs to be more discussion of their application by land users, and their physical and economic feasibility needs to be testing with the new models now available.

- Integration of remote sensing with modelling

Remote sensing data have the potential to make spatially explicit models much more widely applicable. Moreover, if soil moisture and roughness data can be extracted from remote sensing data, we believe the output of the models would be much improved.

- Application of models at national/European scales

  This relates to the last item. If we can gather large European-scale data sets, partly using remote sensing, our models and data have the potential to predict erosion and dust emission at European scales. Similar exercises have been attempted, with some success in the USA.

- Impact analysis of general environmental and agricultural, policies (national or EU) on wind erosion risk by integration of models with socio-economic data

  This relates to the earlier item about modelling. If more economists can be involved, the models could be used to test the economic outcome of particular policies or market changes as they relate to wind erosion and dust emission rates.

- Legislation/policy

  There needs to be more direct two-way contact between the science and the policy.

Discussion

Some more explanation is needed of the Research Priorities. It should be emphasised that:

- They represent a consensus among a large proportion of wind erosion scientists in Europe, although they were developed over only one afternoon (albeit after a three-day conference when these issues were continuously discussed).

- They are priorities for European-level research and policy, not for national or regional research and policy.

- The discussions surrounding the development of the lists and during the meeting had focused on practical issues. Thus the growing awareness of dust as a public health problem (both in Europe and North America) was one of our main concerns. And, following our exposure during the meeting to presentations from North American and Australian contexts, we were aware of the importance of involving farmers in the process of developing the research and policy agendas.

- Finally, we were aware that there had been some excellent research into wind erosion in Europe. We now wished to see it better applied.

European Commission

EUR 20370 — Wind erosion on Agricultural Land in Europe

Luxembourg: Office for Official Publications of the European Communities

2003 — 17.6 x 25 cm

ISBN 92-894-3958-0

Price (excluding VAT) in Luxembourg: EUR 11.50